BTAC

Fundamentals of
POWER ELECTRONICS

Fundamentals of
POWER ELECTRONICS

——————— S. Rama Reddy ———————

CRC Press
Boca Raton London New York Washington, D.C.

Narosa Publishing House
New Delhi Chennai Mumbai Calcutta

Dr. S. Rama Reddy
Professor and Head
Electrical and Electronics Engineering Department
S.R.M. Easwari Engineering College, Chennai-600 089, India

Library of Congress Cataloging-in-Publication Data

Reddy, S. Rama.
    Fundamentals of power electronics/S. Rama Reddy.
      p. cm.
    Includes bibliographical references and index.
    ISBN 0-8493-0934-4 (alk. paper)
     1. Power electronics. I. Title.

TK7881.15.R43 2000
621.381'044-dc21   00-057981

All rights reserved. No part of this publication may be reproduced, stored in a retrieval system or transmitted in any form or by any means, electronic, mechanical, photocopying or otherwise, without the prior permission of the copyright owner.

This book contains information obtained from authentic and highly regarded sources. Reprinted material is quoted with permission, and sources are indicated. Reasonable efforts have been made to publish reliable data and information, but the author and the publisher cannot assume responsibility for the validity of all materials or for the consequences of their use.

Neither this book nor any part may be reproduced or transmitted in any form or by any means, electronic or mechanical, including photocopying, microfilming, and recording, or by any information storage or retrieval system, without prior permission in writing from the publisher.

Exclusive distribution in North America only by CRC Press LLC

Direct all inquiries to CRC Press LLC, 2000 N.W. Corporate Blvd., Boca Raton, Florida 33431. E-mail: orders@crcpress.com

Copyright © 2000 Narosa Publishing House, New Delhi-110 017, India

No claim to original U.S. Government works
International Standard Book Number 0-8493-0934-4
Printed in India

Preface

The present book with simple treatment is the outcome of my many years of teaching. It will serve as an ideal textbook for courses on Power Electronics and Industrial Electronics at polytechnic and undergraduate levels. It will also be very useful to students preparing for professional courses like A.M.I.E. and A.M.I.T.E.

This book is divided into twelve chapters. The first chapter deals with various types of devices, protection, series and parallel operation of SCRs. Chapters 2 and 3 describe various triggering circuits used for SCRs and methods of commutation of SCRs, respectively. Chapter 4 discusses phase controlled converters while Chapter 5 deals with DC to DC choppers. Chapters 6 and 7 deal with various types of inverters and cycloconverters, respectively. Various AC chopper circuits are described in Chapter 8. Chapter 9 gives various applications like speed control of DC and AC motors, heating, welding, UPS, SMPS and HVDC system. Chapter 10 gives analysis of resonant inverters with R and R-L loads. Principle of operation of quasi-resonant converters are given in Chapter 11. Chapter 12 describes microprocessor based triggering schemes for three-phase converter. All the chapters carry short questions and answers at the end.

Appendix I contains basic experiments in power electronics while short question and answer banks are given in Appendix II.

I would like to convey my deep sense of gratitude to Dr. B. Ilango and Dr. V.P. Ramamurthy who taught me this subject.

The author owes his deep gratitude to Mr. T.R. Pachimuthu (Chairman), Dr. T.P. Ganesan (Director) and Prof. V. Subramaniam (Principal), Easwari Engineering College, Chennai for their keen interest in bringing out this book.

The author would like to thank Mr. N. Jayaraman and Mr. N.A. Abbu for drawing the illustrations and Mr. T. Rajasekar for typing the manuscript. Thanks are due to M/s Narosa Publishing House for bringing out this book in its present form.

I would be glad to receive comments and suggestions for the improvement of this book.

S. Rama Reddy

Contents

Preface   *v*

1. Semiconductor Devices   1
   1.1 Introduction *1*
   1.2 P–N Diode *4*
   1.3 Bipolar Junction Transistor (BJT) *6*
   1.4 Silicon Controlled Rectifier (SCR) *7*
   1.5 DIAC *17*
   1.6 TRIAC *17*
   1.7 MOSFET *19*
   1.8 Gate Turn Off Thyristor (GTO Thyristor) *20*
   1.9 Insulated Gate Bipolar Transistor (IGBT) *20*
   1.10 Programmable Unijunction Transistor (PUT) *21*
   1.11 Silicon Controlled Switch (SCS) *21*
   1.12 Silicon Unilateral Switch (SUS) *22*
   1.13 Reverse Conducting Thyristor (RCT) *22*
   1.14 Light Activated SCR (LASCR) *23*
       Short Questions and *Answers* *23*

2. Triggering Circuits   25
   2.1 R-Triggering *25*
   2.2 R-C Triggering *26*
   2.3 UJT Triggering *26*
       Short Questions and Answers *29*

3. Commutation Circuits   30
   3.1 Introduction *30*
   3.2 Commutation *30*
   3.3 Series LC Circuit *30*
   3.4 Ringing Circuit *32*
   3.5 Turn Off Methods *32*
       Short Questions and Answers *37*

4. Phase Controlled Rectifiers   38
   4.1 Introduction *38*
   4.2 Classifications of Rectifiers *38*
   4.3 Performance of Rectifiers *39*
   4.4 Single Phase Rectifiers *42*
   4.5 Half Controlled Rectifier with R-Load *42*

viii   Contents

    4.6   Half Controlled Rectifier with R-L Load   *43*
    4.7   1-φ Fully Controlled Rectifier with R-Load   *43*
    4.8   1-φ Fully Controlled Rectifier (Bridge Type)   *44*
    4.9   1-φ Fully Controlled Rectifier with R-L Load   *45*
    4.10 Fully Controlled Rectifier with Source Inductance   *47*
    4.11a 1-φ Fully Controlled Rectifier with Free Wheeling Diode   *48*
    4.11b Semiconverter   *48*
    4.12 Fully Controlled Rectifier Using One SCR   *50*
    4.13 3-φ Half Controlled Converter (3-Pulse Converter)   *51*
    4.14 3-φ Fully Controlled Bridge Converter (6-Pulse Converter)   *51*
    4.15 Synchronized UJT Triggering Circuit   *58*
    4.16 Firing Circuit for 3–φ Converter   *60*
          Short Questions and Answers   *64*

5. D.C. Choppers   **67**
    5.1   Introduction   *67*
    5.2   Basic Principle   *67*
    5.3   Control Strategies   *68*
    5.4   Classification of Choppers   *69*
    5.5   Voltage Commutated Chopper   *72*
    5.6   Current Commutated Chopper (CCC)   *75*
    5.7   Load Commutated Chopper   *77*
    5.8   Jones Chopper   *78*
    5.9   Step-Up Chopper   *80*
          Short Questions and Answers   *83*

6. Inverters   **85**
    6.1   Introduction   *85*
    6.2   Series Inverter   *85*
    6.3   Parallel Inverter   *87*
    6.4   Bridge Inverters   *88*
    6.5   Mc-Murray Inverter (Voltage Source Inverter)   *90*
    6.6   Mc-Murray-Bedford Inverter   *92*
    6.7   3-φ Inverters   *94*
    6.8   3-φ Current Source Inverter   *94*
    6.9   Voltage Control   *98*
    6.10 Harmonic Control (Waveform Control)   *100*
          Short Questions and Answers   *103*

7. Cycloconverters   **104**
    7.1   Introduction   *104*
    7.2   Single Phase Step Down Cycloconverter   *104*
    7.3   Step-Up Cycloconverter   *105*
    7.4   3-φ to 1-φ Cycloconverter   *105*
    7.5   3-φ to 3-φ Cycloconverter   *106*
    7.6   1-φ to 3-φ Cycloconverters   *108*
          Short Questions and Answers   *108*

Contents ix

8. A.C. Choppers 109
8.1 Introduction *109*
8.2 AC Chopper Using Triac *110*
8.3 AC Chopper with R-Load *111*
8.4 AC Chopper with R-L Load *112*
    Short Questions and Answers *114*

9. Applications 115
9.1 Introduction *115*
9.2 Speed Control of Induction Motor *115*
9.3 Braking of Induction Motor *120*
9.4 Closed Loop Operation *122*
9.5 Speed Control of DC Motors *123*
9.6 Braking of DC Motors *126*
9.7 Closed Loop Controlled DC Drive *126*
9.8a Regulated Power Supply (RPS) *128*
9.8b Switch Mode Power Supply (SMPS) *128*
9.9 Welding *129*
9.10 Heating *129*
9.11 Thyristor Controlled Static On-Load Tap-Changing Gear *131*
9.12 Static Var Compensators *133*
9.13 Generrex Excitation System of Alternators *136*
9.14 Uninterruptible Power Supply *136*
9.15 HVDC Transmission *139*
9.16 Microprocessor Based Thyristor Controlled Electrical Drives *140*
    Short Questions and Answers *146*

10. Resonant Inverters 147
10.1 Static Power Conversion *147*
10.2 Advantages of Zero Voltage Switching *147*
10.3 Disadvantages of Zero Voltage Switching *148*
10.4 Resonant DC Link Converter *148*
10.5 Base Drive Circuit of Bipolar Junction Transistor *149*
10.6 Single Phase IPMRI *150*
10.7 Analysis of IPMRI with R-Load *154*
10.8 Single Phase IPMRI with R-L Load *154*

11. Quasi Resonant Converters 157
11.1 General *157*
11.2 Zero Current Switching *157*
11.3 Resonant Switch Topologies *158*
11.4 Principle of Operation of QRC *159*

12. Microprocessor Based Triggering Schemes 163
12.1 Firing Scheme for 3 Phase Converters Proposed by Huy, Roye and Perret *163*
12.2 Firing Scheme Proposed by S.B. Dewan *165*

Appendix I: Basic Experiments in Power Electronics — 168
1. SCR Characteristics *168*
2. UJT Firing Circuit *170*
3. Triac Characteristics *171*
4. RC Triggering Scheme of SCR *172*
5. Voltage Commutation *173*
6. Current Commutation *176*

Appendix II: Short Questions and Answers — 178
Questions and Answers Bank-I *178*
Questions and Answers Bank-II *185*
Applications *187*

REFERENCES — 188
INDEX — 189

1
Semiconductor Devices

1.1 INTRODUCTION
In this chapter, types of semiconductors, various devices and protection of SCR circuits are discussed.

1.1.1 Intrinsic Semiconductor
Intrinsic semiconductor is a pure semiconductor. Silicon and germanium are the semiconductor materials. To form a stable covalent bond, 8 valence electrons are required. The silicon atom at the centre has 4 valence electrons. It shares 4 electrons from the neighbour atoms to form a covalent bond. The symbol of the covalent bond is shown in Fig. 1.1(a).

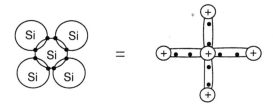

(a) Covalent bond

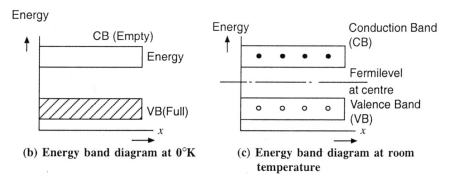

(b) Energy band diagram at 0°K   (c) Energy band diagram at room temperature

Fig. 1.1

At absolute zero temperature, no energy is supplied to the crystal. All the electrons are engaged in forming covalent bond and no free electrons are available. Hence there is no conduction. Thus the semiconductor acts as an insulator at 0°K. The Energy band diagram for this condition is shown in Fig. 1.1(b). The conduction band is empty as no conduction electrons are available.

When thermal energy is supplied to the semiconductor, some of the covalent bonds are broken due to the energy supplied. These electrons jump from valence band to the conduction band as shown in Fig. 1.1(c).

1.1.2 Extrinsic Semiconductor

Extrinsic semiconductor is also called impure semiconductor. They are classified as N-type and P-type. N stands for negative and P stands for positive.

(a) Negative Type or N-Type

When the intrinsic semiconductor is doped with pentavalent impurity, negative type semiconductor is formed. The pentavalent impurities are antimony, arsenic and bismuth. The pentavalent atom at the centre has 5 valence electrons. This atom shares four electrons from the neighbour atoms. For the formation of stable covalent bond, only 8 electrons need to rotate in the valance orbit. Thus one excess electron is produced by each impurity atom. Several impurity atoms donate several electrons. Since the impurity atoms donate electrons, they are known as *donors*. The energy band diagram is shown in Fig. 1.2(b). A few covalent bonds are broken at room temperature due to the thermal energy supplied by the nature. The vacancies are shown as holes in the valence band. The majority carriers are electrons and the minority carriers are holes.

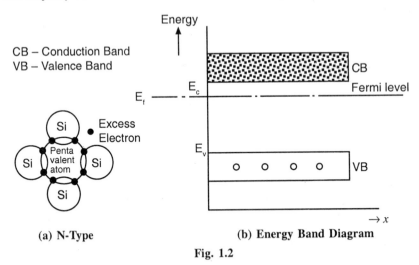

(a) N-Type (b) Energy Band Diagram

Fig. 1.2

E_c is the lowest energy level of the conduction band and E_v is the highest energy level of the valence band. E_f is the Fermi Energy level. Fermi level corresponds to the centre of gravity of the electrons and holes. In the case of intrinsic semiconductor, the number of electrons are equal to the number of holes. The Fermi level lies midway between the valence band and conduction band. In the N-type semiconductor the fermi level is lifted towards the conduction band as the conduction electrons are the majority carriers.

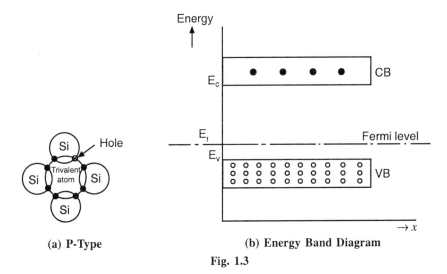

(a) P-Type (b) Energy Band Diagram
Fig. 1.3

(b) Positive Type or P-Type

When the intrinsic semiconductor is doped with trivalent atoms, positive type semiconductor is formed. The trivalent atoms are indium, gallium, boron and aluminum. The trivalent atom at the centre has 3 valence electrons. This atom shares 4 electrons from the neighbour atoms. 8 electrons are required to form the valence orbit. In other words the trivalent atom can accept one electron. This vacancy is known as a *hole*. The holes have positive charge. Millions of impurity atoms can accept millions of electrons. Hence they are called as *acceptors*. The majority carries are holes and the minority carriers are electrons.

The electrons jumped from the valence band to the conduction band due to thermal energy are represented in the conduction band. Valence orbit of each impurity atom has one hole. Thus, holes in the valence orbits of the impurity atoms are represented in the valence band as shown in the energy band diagram. The Fermi level is shifted down as the majority carriers are the holes in the valence orbits.

1.1.3 P-N Junction

Holes are represented by +ve sign and the impurity atoms by –ve ions in P-type semiconductor. Electrons are represented by –ve sign and the impurity atoms by +ve ions in N-type semiconductor.

When the P-N junction is formed, the conduction electrons in the N-region will diffuse (penetrate or enter) into the P-region. The holes in the P-region will diffuse into the N-region. The electrons fall into the holes. This process is known as Recombination. The average time for which the electron travels before it recombines, is known as the Life Time. A depletion layer is formed at the junction. The recombination stops after some time. The conduction electrons on the N-side are repelled by the negative ions on the P-side. The holes on the P-side are repelled by the positive ions on the N-side. Thus a restraining force is set up at the junction and this prevents further recombination.

4 *Fundamentals of Power Electronics*

In the graph, the charge density on the left side of the potential barrier is negative and it is positive on the right side of the potential barrier as indicated in the charge density curve.

The Electric Field Intensity is maximum at the centre and the magnitude of electric field Intensity decreases on either side. Potential is the work done in moving a unit positive charge from the left to the right side of depletion layer. The work done has to be increased as the charge is moved to the right side. Therefore the potential value increases as we proceed to the right side.

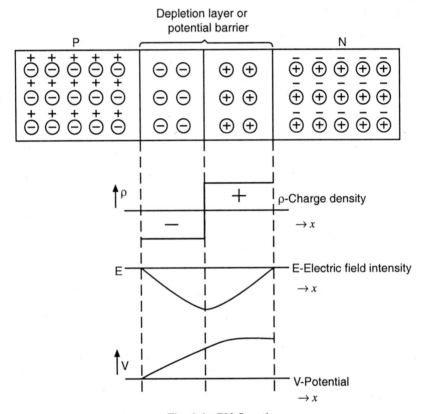

Fig. 1.4 PN Junction

(a) Depletion Layer
When the P-N junction is formed, the electrons and holes move towards the junction. The electrons will fall into the holes. At the junction, a charge free region is formed. The positive ions and the negative ions are separated by a small distance. These are nothing but the dipoles. The direction of the Electric field is given by the force experienced by the unit positive charge kept in the electric field. The direction is away from the positive charge. An electric field at this layer implies that there is a potential due to the separation of the charges. This potential is known as the barrier potential. This is equal to 0.3 V for germanium and 0.7 V for Silicon.

1.2 PN-DIODE

Diode is a two layer device as shown in Fig. 1.5(a). The junction is formed using P and N layers. The two electrodes of the diode are anode and cathode. When positive terminal of the battery is connected to the anode and negative terminal to the cathode, the diode is said to be forward biased. A stream of electrons start from the negative terminal of the battery and they flow through the N layer as conduction electrons. Near the junction, electrons fall into the holes and they become valance electrons. They travel through the P-layer as valance electrons and later these electrons are attracted by the positive terminal of the battery. This explains the conduction of diode when it is forward biased. From the forward characteristic it can be seen that there is no current till the applied voltage is less than the knee voltage (V_k). When the voltage is more than the V_k, the current through the diode increases with the increase in the applied voltage as shown in Fig. 1.5(b).

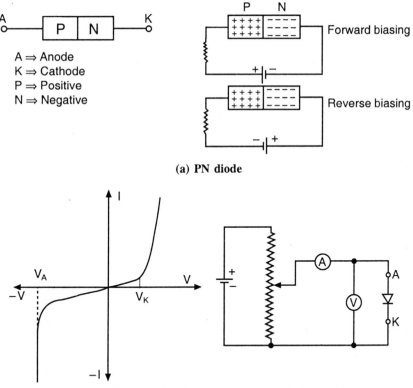

A ⇒ Anode
K ⇒ Cathode
P ⇒ Positive
N ⇒ Negative

(a) PN diode

V_K = Knee voltage (0.3 for Germanium and 0.7 for Silicon)
V_A = Avalanche breakdown voltage

(b) Characteristics of diode

Fig. 1.5

When the anode is connected to the negative terminal and cathode to the positive terminal of the battery, the diode is said to be reverse biased. Current of the order of micro Amperes flows due to the minority carriers.

6 *Fundamentals of Power Electronics*

When the voltage is increased beyond V_A, electric field increases. The electron entering this field experiences more force and this electron can knock off another electron from the covalent bond. Again these two can knock off another two and this process continues. A large current flows through the diode. This effect is called avalanche effect. If the current is not limited, the device gets damaged.

The diode can be represented as a closed switch when it is forward biased and a open switch when it is reverse biased.

Signal diodes are rated for low voltage and low current. The wattage will be 1/2 watt or 1 watt. Power diodes are rated at high voltage and high current. They are of the order of Kilovolts and Kiloamperes. The frequency of operation of the signal diodes will be higher than that of power diodes.

1.3 BIPOLAR JUNCTION TRANSISTOR (BJT)

Transistor is a three layer device having two Junctions. Three terminals are emitter, base and collector. The principle of transistor operation can be explained with Fig. 1.6(a). It can be seen that the base is very thin. Let 100% of electrons start from the negative terminal of the input battery. These electrons flow through the emitter. In the base region, about 1% of conduction electrons fall into the holes and the remaining 99% flow through the collector terminal. These electrons are attracted by the positive terminal of the Output battery, thus

$$I_e = I_b + I_c$$

In common base configuration, the base terminal is used as the common terminal for input and output. The gain $\alpha = I_c/I_e$. This is approximately equal to 1. In common emitter configuration, emitter terminal is grounded. The gain $\beta = I_c/I_b$. This varies between 10 to 250.

The output characteristic can be plotted with the observations of circuit shown in Fig. 1.6(b). For a fixed base current the variation of output current with the variation in the output voltage is noted. Similarly, readings are obtained for various values of base currents and the output characteristics are plotted. From the characteristic shown in Fig. 1.6(b), it can be seen that the current is negligible in the cut off region. In the saturation region, the voltage across the transistor is very small (about 1 V). This state is ON state.

Transistor is a current controlled device. When base drive is given, the transistor conducts, the entire voltage drops across R_c and $V_{CE} = 0$. When there

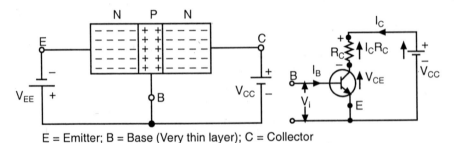

E = Emitter; B = Base (Very thin layer); C = Collector

(a) Bipolar junction transistor

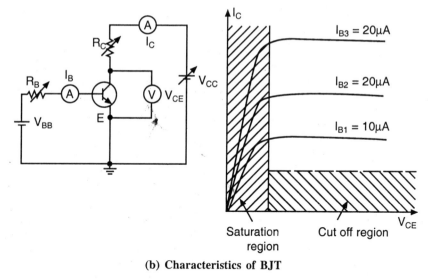

(b) Characteristics of BJT

Fig. 1.6

is no base drive, the transistor does not conduct and $V_{CE} = V_{cc}$. Thus transistor act as a controlled switch.

When the transistor is saturated, it acts as a closed switch and when it is in the cut off region, it acts as an open switch.

If a pulse or base drive is given at the base, the transistor acts as a closed switch. If the pulse is not given, it acts as open switch. This concept is very useful in understanding power electronic circuits.

1.4 SILICON CONTROLLED RECTIFIER (SCR)

SCR is a four layer device having three junctions namely J_1, J_2, and J_3. When the SCR is forward biased, the junctions J_1 and J_3 act as closed switches since they are forward biased. J_2 acts as a open switch since it is reverse biased. There cannot be a current from anode to cathode. The SCR is said to be in forward blocking state. The configuration and characteristic are shown in Fig. 1.7(a) and 1.7(b) respectively.

If the voltage is increased beyond V_{FB}, the junction J_2 breaks down and the SCR goes to conduction state. The voltage across the device reduces to 1 V since most of the applied voltage drops across the resistance in the anode circuit.

When the anode of the SCR is connected to the negative terminal and the cathode to the positive terminal of the battery, the junctions J_1 and J_3 are reverse biased and J_2 is forward biased. There cannot be a current from cathode to anode. The SCR is said to be in reverse blocking state. The equivalent circuit of the SCR in the reverse bias condition has two diodes in series with reverse voltage being applied. Therefore the characteristic of the thyristor in the reverse biased condition is similar to that of a diode.

The process of changing the SCR from OFF state to ON state is called turn on. The process of bringing the SCR from ON state to OFF state is called turn-off.

8 Fundamentals of Power Electronics

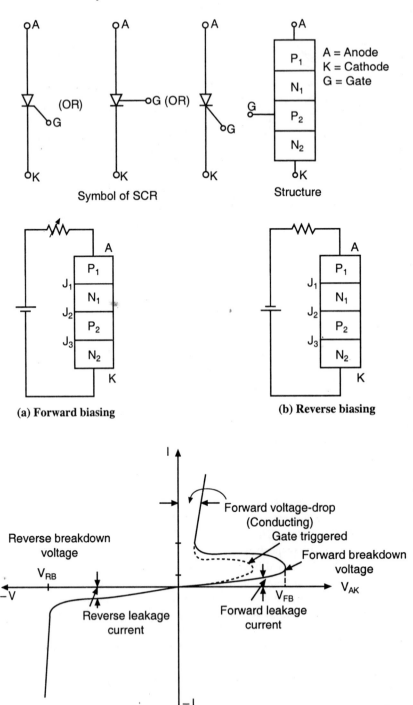

(a) Forward biasing

(b) Reverse biasing

(c) Characteristics of SCR

Fig. 1.7

1.4.1 Two Transistor Analogy of SCR

The two transistor analogy of SCR is shown in Fig 1.8(a). When gate is made positive with respect to cathode, the NPN transistor conducts and this provides base current for the PNP transistor. The output of PNP is the input to the NPN transistor. Even though the gate voltage is removed the transistors continue to conduct since they are connected back to back. The two transistors finally get saturated and thus the SCR acts as a closed switch when it is forward biased and a positive pulse is applied at the gate. Without gate voltage (pulse), a large voltage has to be applied between anode and cathode. With a gate pulse, a small voltage is sufficient to make the SCR conduct. The arrangement shown in Fig. 1.8b is similar to an ideal motor generator set which can run without external supply. The two machines are connected back to back.

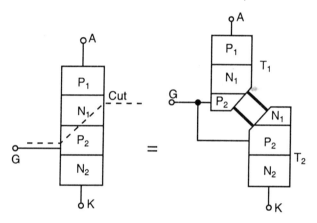

Fig. 1.8(a) Two transistor analogy

The anode voltage required for conduction is a function of the gate current. From the characteristic, it can be seen that larger the value of gate current, smaller the anode voltage required for conduction.

For transistor (1),

$$I_{E1} = I_{C1} + I_{B1}$$

Dividing the above throughout by I_{E1}, we get

$$1 = \frac{I_{C1}}{I_{E1}} + \frac{I_{B1}}{I_{E1}}$$

Substitute $\alpha_1 = \dfrac{I_{C1}}{I_{E1}}$

$$1 = \alpha_1 + \frac{I_{B1}}{I_{E1}} \quad \text{or} \quad 1 - \alpha_1 = \frac{I_{B1}}{I_{E1}}$$

that is

$$I_{B1} = (1 - \alpha_1) I_{E1},$$

From Fig. 1.8(b) $I_{B1} = I_{C2} = (1 - \alpha_1) I_A$ \hfill (1.1)

10 Fundamentals of Power Electronics

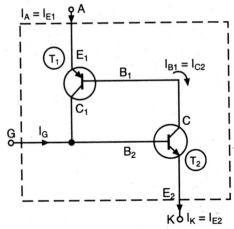

Fig. 1.8(b)

For transistor (2)

$$\alpha_2 = I_{C2}/I_{E2}; \quad I_{C2} = \alpha_2 I_{E2}$$

From the Fig. 1.8(b) $I_{E2} = I_K$

Therefore
$$I_{C2} = \alpha_2 I_K \qquad (1.2)$$

Equations (1.1) and (1.2) are equal.

Therefore
$$(1 - \alpha_1) I_A = \alpha_2 I_K \qquad (1.3)$$

From the figure, by KCL

$$I_A + I_G = I_K \qquad (1.4)$$

Substitute (1.4) into (1.3)

$$(1 - \alpha_1) I_A = \alpha_2 (I_A + I_G); \quad I_A (1 - \alpha_1 - \alpha_2) = \alpha_2 I_G$$

Therefore $I_A = \alpha_2 I_G [1 - (\alpha_1 + \alpha_2)]$

$$\frac{I_A}{I_G} = \frac{\alpha_2}{1 - (\alpha_1 + \alpha_2)} \qquad (1.5)$$

If $\alpha_1 + \alpha_2 = 1$, then I_A/I_G tends to ∞. Practically $\alpha_1 + \alpha_2$ is around 0.99. From the equation (1.5), it can be seen that using SCR, very large anode current can be controlled by using small gate current since the ratio of I_A to I_G is very high.

Let us consider that the SCR shown in Fig. 1.7(a) to be in on state. The anode current can be reduced by increasing the resistance in the anode circuit. As the resistance in the anode circuit increases, the anode current decreases. Below a particular current value (I_h), the SCR fails to remain in the ON state. The minimum current below which SCR goes to OFF state is called holding current. Typical value is 20 mA. Let us consider the SCR to be in the OFF state. The minimum current required to keep the SCR latched after it is turned on and the gate signal is removed is called latching current.

1.4.2 Reverse Bias Condition of SCR

When an SCR is reverse biased, positive voltage should not be applied at the gate. The junctions J_1 and J_3 are reverse biased and J_2 is forward biased. When gate is made positive with respect to cathode, the junction J_3 is also forward biased. Hence the entire voltage appears across J_1. The junction J_1 might breakdown by avalanche breakdown principle. There will be a current from the cathode to anode. But the SCR is supposed to conduct only when it is forward biased and positive voltage is applied at the gate. Hence the SCR gate should not be made positive when it is reverse biased.

1.4.3 Difference Between Transistor and SCR

Transistor acts as a controlled switch. It acts as a closed switch as long as base drive is available at the input. It acts as an open switch when there is no base drive. To keep the transistor in ON state, a continuous base drive is needed at the input. In the case of SCR, a continuous dc voltage is not required at the gate to keep the SCR on. Only a pulse is needed at the gate since the SCR is a self latching device.

1.4.4 Ratings of Thyristor

(a) Peak Inverse Voltage

It is defined as the maximum voltage which the device can safely withstand in

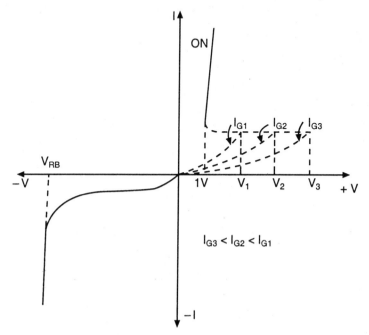

Fig. 1.9 Effect of gate current

its OFF state. The voltage withstanding capacity of a thyristor is temperature dependent.

(b) On-State Voltage
When a thyristor starts conducting, the voltage across the device is very small. The voltage which appears across the device during its ON state is known as its On state voltage. The normal value of the voltage varies from 1 V to about 4 V.

(c) Rate of Rise of Voltage (dv/dt)
The rate at which the voltage across the device rises without triggering the device, is known as its rate of rise of voltage. The maximum rate of rise of voltage without triggering the device is called its critical rate of rise of voltage. As soon as the rate of rise of voltage increases more than the critical rate of rise of voltage, the device starts conducting. The depletion layer in a thyristor forms a capacitor of capacitance, say C. If I_c is the capacitive current flowing through the junction in the device at any moment, then $I_c = C\, dv/dt$, where dv/dt is its rate of rise voltage.

(d) Current rating
The current carrying capacity of the device is known as its current rating. It can be of two types viz (i) continuous and (ii) intermittent.
(i) *Continuous rating:* It is normally specified in terms of average or r.m.s. value.
(ii) *Repetitive rating:* These are given in terms of peak value. When the SCR is made ON-OFF continuously, it is subjected to repeated transients. These ratings are then to be taken care of.
(iii) *Non-repetitive rating:* These are also called surge ratings. These ratings are given in terms of peak value. The device can be subjected to surge voltages or currents only once in a while during its operating life. Surge ratings correspond to the maximum possible non-repetitive voltage or current peak that the SCR can withstand. During a surge, the junction temperature may exceed the permissible level for a short duration. However as long as the magnitude of such surges is less than the specified value, the device regains its normal temperature as soon as the spike terminates. This will not give rise to any faulty operation.

(e) Latching Current (I_l)
It is that minimum value of current which is required to latch the device from its OFF-state to its On-state. It can also be defined as the minimum value of current required to trigger the device. Its value is normally in milli amperes (mA).

(f) Holding Current (I_H)
It is the minimum value of current required to hold the device in conduction state. It may be defined as minimum value of current, below which the device stops conducting and returns to its OFF state. The value of this current is very small (usually in milliamperes) and even lesser than that of the latching current of the device.

(g) Gate Current (I_g)

The current which is applied to the gate of the device for control purposes, is known as its gate current. It may be of two types viz., (i) the minimum gate current ($I_{g\ min}$), (ii) and maximum gate current ($I_{g\ max}$). The minimum gate current is the minimum value of current required at the gate of triggering the device. A current below $I_{g\ min}$ will not be able to trigger the device. The value of $I_{g\ min}$ depends on the rate of rise of current. On the other hand, the maximum gate current ($I_{g\ max}$) is the maximum value of current which can be applied to the device safely. Current higher than this, if applied, may damage the gate.

(h) Gate Power Loss (P_g)

The mean power loss which occurs because of the flow of gate current between the gate and the main terminals, is known as the gate power loss (P_g).

(i) Turn ON Time (T_{ON})

Though the thyristor is an extremely fast switching device, it does not conduct instantaneously. It takes finite time to reach its full conduction state after being triggered. The time for which the device waits before achieving its full conduction is known as its Turn-ON time.

(j) Turn-OFF Time (T_{OFF})

A reverse voltage is required to be applied across the device for switching it OFF. After applying the reverse voltage, it takes however small, a finite time to get switched OFF (commutated). This time is defined as the Turn-OFF time of the device. Its usual value is around 100 μsec for converter grade SCRs and 20 μsec for inverter grade SCRs.

(k) Rate of Rise of Current (di/dt)

The rate at which the current flowing in the device rises is known as its rate of rise di/dt. The maximum rate of change of current which the device can withstand in its On state, is called its critical rate of rise of current.

1.4.5 Protection of SCR Circuits

For the reliable operation of SCR circuits, the protection of SCR is essential. Two major protections are over current protection and over voltage protection. The over current protection in Fig 1.10(a) is provided using fuses at the input and load. When over current flows due to any fault, the fuse wire gets heated and it blows off. The over voltage protection is provided by a thyractor. It is a nonlinear resistor. It offers very high resistance at rated voltage. When impulse voltage reaches the circuit, it offers very low resistance. Therefore the impulse voltage does not reach transformer, SCR and load.

The SCR also needs dv/dt and di/dt protection. The di/dt protection is provided by using a small inductor in series with the SCR. The dv/dt protection is provided by using a snubber circuit. A snubber circuit has a resistance and capacitance in series. In the Fig. 1.10(b), if the switch S is closed without snubber, a large current flows through the SCR due to switching transient. If snubber is present, this current flows through the snubber since the capacitance offers very low reactance during transient. The resistance R limits the current through the snubber circuit when SCR conducts. Snubber acts as a dv/dt shock absorber.

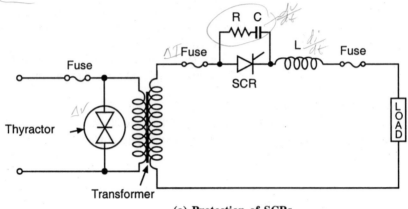

(a) Protection of SCRs

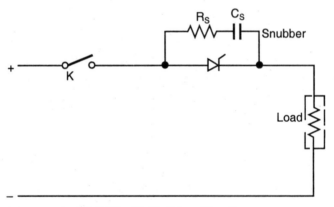

(b) SCR with Snubber Circuit

Fig. 1.10

1.4.6 Crow Bar Protection

The SCR connected across the supply in Fig. 1.11 is called crow bar SCR. Under normal operating conditions the current is less than the rated current; the voltage across the resistor is less since the current is less. The control circuit does not generate a pulse when the voltage is less. Under fault conditions, large current flows through the load and more voltage is available across the resistance. The control circuit generates a pulse and the SCR is turned ON. When the SCR conducts, a large current flows through the fuse and it immediately blows off. Thus the crow bar SCR and control circuit protects the load from over currents.

1.4.7 Series and Parallel Operation of SCRs

In HVDC transmission systems, SCRs are used for rectification. The operating voltage is 800 K.V. To meet this voltage, several SCRs are connected in series.

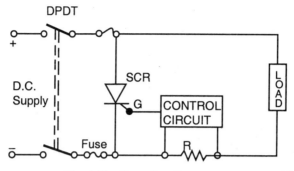

Fig. 1.11 Crow Bar Protection

From the characteristic, it can be observed that the voltages across the devices are not equal for the same current. The voltage across the devices can be equalised by connecting high resistance in parallel with each thyristor. Thus several SCRs are operated in series by using resistances in parallel with them as shown in Fig. 1.12(a). The value of this resistance must be as high as possible.

Furnaces, large capacity A.C. motors and D.C. motors operate at very high currents. This large current can be handled by connecting more SCRs in parallel. The characteristics of two SCRs are shown in the Fig. 1.12(b). It can be seen that the currents shared by the SCRs are not equal. This is similar to two brothers of a family not sharing the load equally. This problem can be solved by having a centre tapped reactor. When the current through T_1 tries to increase, an e.m.f is induced in L, in such a way to reduce the current through T_1 and the e.m.f across L_2 acts in such a way to increase the current through T_2. Thus the centre tapped reactor tries to maintain equal currents through the SCRs. Parallel operation with more number of thyristors is not popular due to the difficulty in the design of the reactor.

If the turn-on time of $T_1 < T_2$, then T_1 turns on first and the voltage reduces to one volt. Later T_2 cannot turn on since the voltage is reduced to 1 V.

1.4.8 Methods of Turn-On
Various methods of turn-on are
1. Increasing forward voltage.
2. Gate turn on
3. dv/dt turn on
4. Light turn on
5. Thermal turn on

1. Increasing Forward Voltage
When the SCR is forward biased, J_1 and J_3 act as closed switches and J_2 acts as an open switch since it is reverse biased. When the applied voltage is increased beyond forward break over voltage, J_2 breaks down and a current flows from anode to cathode. This method is not generally used since the required voltage is large.

SERIES AND PARALLEL OPERATION OF SCRs

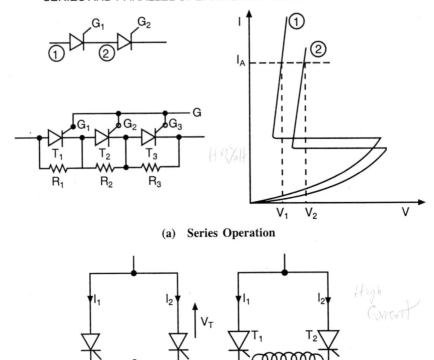

(a) Series Operation

(b) Parallel Operation

Fig. 1.12

2. GATE turn on
Using two transistor analogy, it can be seen that conduction can be achieved with a pulse at the gate. Both transistors conduct and the SCR acts as a closed switch. In the gate circuit, the source V_2 injects electrons from cathode to gate. The junction J_2 can easily break down since V_2 supplies large number of conduction electrons. This method is the most popular method of turning on the SCR.

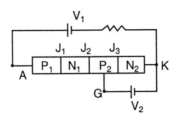

3. dv/dt turn on
When anode is made positive with respect to cathode, the junctions J_1 and J_3 are forward biased and J_2 is reverse biased. The depletion layer around J_2 acts

as a dielectric. The junctions J_1 and J_3 act as good conductors. The SCR now behaves like a capacitor. When the switch is suddenly closed dv/dt is large. Large charging current [$c\ dv/dt$] flows through the SCR and the junction J_2 breaks down. Therefore the SCR turns on. This method is not generally used since large charging currents may damage the device.

4. Light turn on

In this method the energy required for breaking covalent bonds is obtained from light energy. The energy required for breaking the covalent bonds is given by a source of light. A few covalent bonds break and the process continues. The junction J_2 breaks down and the SCR gets turned on.

In HVDC systems an isolation is required between the power circuit and the control circuit. This is inherently available with light activated SCRs (LASCR).

5. Thermal turn on

In this method, the energy required for breaking the covalent bonds is obtained from heat energy. The area around J_2 is heated by means of an external thermal source. The junction J_2 breaks down and the current flows from anode to cathode.

1.5 DIAC

Diac is a five layer four junction device. This is shown in Fig. 1.13. The word diac can be split into DI and AC. DI stands for two electrodes namely M_{T1} and M_{T2}. AC indicates its ability to conduct in both the directions. From the structure, the equivalent circuit can be drawn with two PNPN devices connected back to back. When M_{T2} is made positive with respect to M_{T1}, the device 1 is forward biased. Junctions J_2 and J_4 are forward biased and J_3 is reverse biased. When the voltage is increased beyond the break over voltage the junction J_3 breaks down and the device 1 goes from high impedance state to low impedance state. The characteristic is similar to that of SCR forward characteristic without gate facility.

When M_{T1} is made positive with respect to M_{T2}, the device 2 is forward biased. The junctions J_1 and J_3 are forward biased and J_2 is reverse biased. J_2 breaks down when the voltage is increased beyond break over voltage and the current flows from M_{T1} to M_{T2}. The characteristic in the third quadrant is similar to that of first quadrant. The diac is used in the triggering circuit of triac.

1.6 TRIAC

The equivalent circuit and characteristic of triac are shown in Fig. 1.14(a) and 1.14(b) respectively. Triac is having three terminals namely M_{T1}, M_{T2} and gate (G). The word triac can be split into Tri and ac. Tri stands for three and ac for its ability to conduct in the both directions. The triac is equivalent to two thyristors connected back to back with their gate terminals tied up. When M_{T2} is positive with respect to M_{T1}, the SCR 1 is forward biased. If the gate is made positive with respect to M_{T1}, the SCR 1 conducts and the device goes from high impedance state to low impedance state. When M_{T1} is made positive with respect to M_{T2}, SCR 2 is forward biased and it conducts when the gate is made positive. Thus the triac can conduct in both the directions. The SCR demands

18 *Fundamentals of Power Electronics*

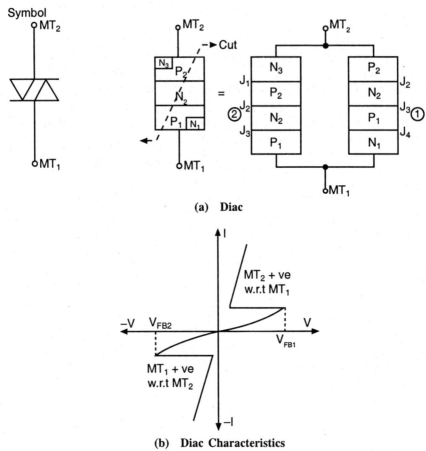

(a) Diac

(b) Diac Characteristics

Fig. 1.13

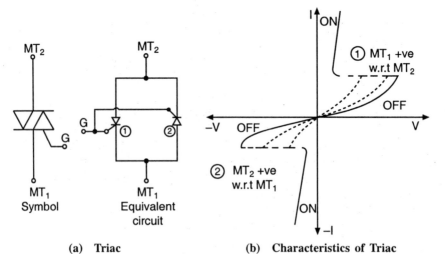

(a) Triac   (b) Characteristics of Triac

Fig. 1.14

a positive voltage between gate and cathode. But the triac can conduct with either positive or negative voltage at the gate.

1.7 MOSFET

Metal oxide semiconductor field effect transistor (MOSEFT) has three terminals namely drain, source and gate as shown in Fig. 1.15(a). The terminology comes from the overhead tank system having source, drain and a gate valve to control the flow of water from source to drain. Electrons from source to drain are controlled by varying the potential of gate. Between the gate terminal and NPN device, oxide coating is provided.

Gate acts as one plate, the PN device acts as another plate and the oxide acts as a dielectric. When the source V_g is connected to the gate, a positive charge is available at the gate. By the principle of parallel plate capacitor an equal amount of negative charge is available in the PN device. The p-layer looses its identity because it is filled with lot of conduction band electrons. The junction

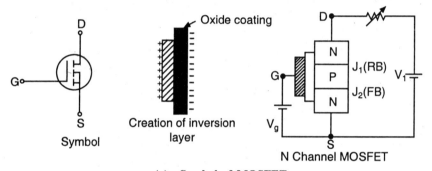

(a) Symbol of MOSFET

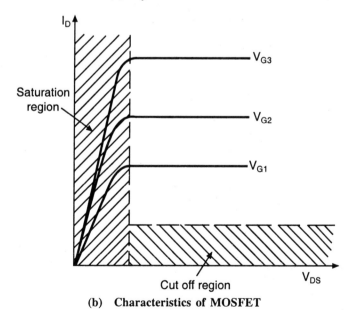

(b) Characteristics of MOSFET

Fig. 1.15

J_1 easily breaks down and current flows from drain to source. When the gate is made positive it conducts. Otherwise it does not conduct. Thus the MOSEFT acts as a voltage controlled device. MOSFET is used at high switching frequencies (10 kHz to 1 Mz).

When a MOSFET is used with positive voltage at gate, it is called enhancement only MOSFET. From the characteristic, it can be seen that the drain current increases with the increase in the gate voltage.

1.8 GATE TURN OFF THYRISTOR (GTO THYRISTOR)

The SCR cannot be turned off using the gate. GTO can be turned off by using the gate. The bidirectional arrow in the symbol indicates that the gate current can flow in both the directions. To turn off of the GTO, a gate current of about 1/5 of the anode current is required. The turn on and turn off times of GTO are less than those of SCR.

When a negative voltage is applied between gate and cathode, the excess charges in the NPN transistor are attracted by the battery. The collector current of T_2 reduces. In other words the base current of T_1 reduces. Collector current of T_1 reduces. This will reduce base current of T_2. This continues until the two transistors are turned off.

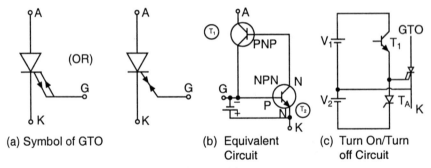

(a) Symbol of GTO (b) Equivalent Circuit (c) Turn On/Turn off Circuit

Fig. 1.16

The turn on cum turn off circuit is shown in Fig. 1.16c. When the transistor T_1 is turned on, the source V_1 is connected between gate and cathode. Gate of GTO is made positive with respect to cathode. GTO turns on.

To turn off GTO, auxiliary SCR T_A is turned on. The source V_2 is connected between gate and cathode. Gate is made negative with respect to cathode and GTO gets turned off. The characteristic of GTO is similar to that of SCR.

1.9 INSULATED GATE BIPOLAR TRANSISTOR (IGBT)

IGBT has the combined characteristics of BJT and MOSFET. It has high input impedance and fast turn on and turn off like MOSFET. It has high power handling capability like BJT.

The function of IGBT is same as that of BJT and MOSFET. IGBT acts as a switch. Output characteristics of IGBT are similar to that of BJT.

The turn on time of IGBT is less than that of BJT. The IGBT needs reverse bias during turn off. The on state value of V_{CE} for IGBT is slightly higher than that of BJT.

1.10 PROGRAMMABLE UNIJUNCTION TRANSISTOR (PUT)

PUT has a structure similar to that of a SCR. The gate terminal is taken fron N_1 instead of P_2. This gate is called "Anode gate". The characteristics of PUT are similar to that of SCR. Its largest rating is 200 V and 1 A.

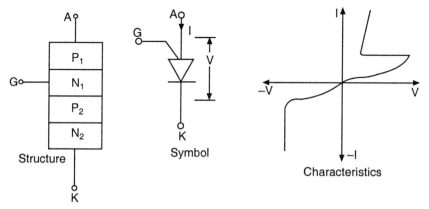

Fig. 1.17 Programmable Unijunction Transistor

1.11 SILICON CONTROLLED SWITCH (SCS)

SCS is a four layer device with two gates namely cathode gate and anode gate. This device is similar to a vacuum device tetrode. The characteristic of SCS are similar to the characteristic of SCR.

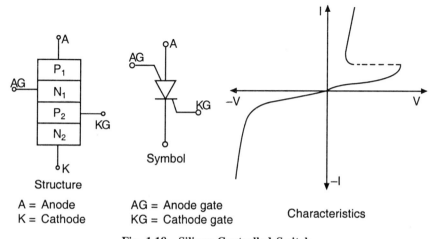

A = Anode   AG = Anode gate
K = Cathode   KG = Cathode gate

Fig. 1.18 Silicon Controlled Switch

SCS can be turned on by either gate. A large reverse current through anode gate can be used to turn it off. Its rating is around 100 V and 200 mA. This is used for timing, logic and triggering circuits.

1.12 SILICON UNILATERAL SWITCH (SUS)

A SUS is similar to a "PUT" but with an in built low voltage avalanche diode between gate and cathode as shown. Because of the presence of diode, SUS turns on for a fixed anode to cathode voltage unlike an SCR whose trigger voltage and current vary widely with changes in ambient temperature. SUS is used mainly in timing, logic and trigger circuits. Its rating is about 20 V, 0.5 A.

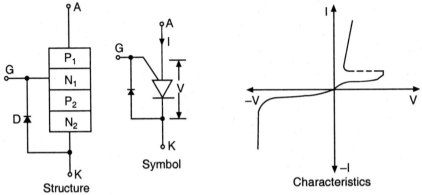

Fig. 1.19 Silicon Unilateral Switch

1.13 REVERSE CONDUCTING THYRISTOR (RCT)

In chopper and inverter circuits, diode is connected in antiparallel (i.e back to back) with the SCR. This diode is called feed back diode. RCT is a single packing in which both SCR and diode are present. The forward characteristic of RCT is same as that of SCR. If the RCT is reverse biased, the internal diode gets forward biased and it starts conducting. Therefore the reverse characteristic of RCT is same as the forward characteristic of a diode.

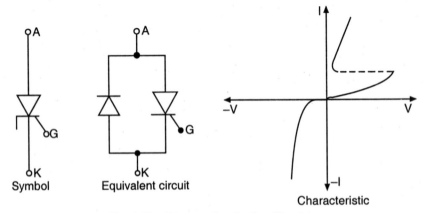

Fig. 1.20 Reverse Conducting Thyristor

RCT is also called "asymmetrical thyristor" (ASCR). The forward blocking voltage varies from 400 to 200 V and the current rating goes up to 500 A. The reverse blocking voltage is typically 30 to 40 V.

1.14 LIGHT ACTIVATED SCR (LASCR)

In this SCR, the gate is turned on by using light energy. It can be seen that the isolation exist between the power circuit and control circuit. Whenever the SCR has to be turned on, the control circuit generates a pulse. This pulse makes the LED to conduct. When the LED conducts the light falls on the gate surface of the SCR. The SCR gets turned on. LASCR is used in H.V.D.C. (High Voltage Direct Current) Transmission system.

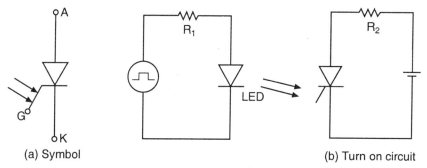

(a) Symbol   (b) Turn on circuit

Fig. 1.21   LASCR

Short Questions and Answers

1. With the increase in gate current, the breakdown voltage of an SCR *decreases*.
2. In a UJT, what is the relation between V_P and V_{BB}?

$$\text{Intrinsic stand off ratio } \eta = \frac{V_P}{V_{BB}} = \frac{V_{EB2}}{V_{BB}}$$

3. No. of the thyristors connected in parallel provide total rated current *less than* the sum of individual ratings.
4. Inverter grade SCRs have t_q less than *25 μs*.
5. Excess dv/dt to a thyristor may cause *mal operation*.
6. In case of thyristor in series, both forward and reverse voltages have to be *shared*.
7. The collector current is 2.9 mA in a certain transistor. If the base current is 100 μA, what is the value of α.

$$\text{Ans. } \alpha = \frac{I_c}{I_E} = \frac{I_c}{I_c + I_B} = \frac{2.9}{3} = 0.97$$

8. An example of solid state device is *FET*.
9. GTO can be turned-off by applying *large* negative current.
10. Turn-OFF time of SCR depends on *temperature* and *forward current*.
11. What is a PN junction?

    Ans. If a pure semiconductor is partly doped withh 3rd group and partly with fifth group impurity, a PN junction is formed. If offers low resistance in forward direction and high resistance in reverse direction.

24 *Fundamentals of Power Electronics*

12. What are the members of thyristor family?

    Ans. SCR, Diac, Triac, SUS, GTO and LASCR.

13. What is the effect of negative gate current on a normal SCR?
    Ans. Gate has no control over the turn-off of the device. Negative gate current produces additional heat at the gate junction.

14. How the forced turn-off of an SCR is different from the natural turn-off?
    Ans. In AC circuits, the current naturally reduces to zero. In DC circuits the current is forced to zero by using L and C.

15. The reverse recovery time of a diode is the time required by *carriers* to recombine and get neutralised.

16. Compared to that of the thyristor, the switching loss of GTO is much lower since turn-ON and turn OFF times of GTO are less.

17. A transformer is employed in the triggering circuit for *isolation* purpose.

18. A diode is usually connected across the primary of a pulse transformer to prevent dc *saturation*.

19. UJT is most suitable for being used in an oscillator circuit due to existence of *negative* resistance region.

20. Ferrite core is used in pulse transformer to ensure no *dc saturation*.

21. The most suitable Gate signal for SCR is *high frequency pulse train*.

22. Is the latching current more than holding current?

    Ans. Yes.

23. What is the difference between holding and latching currents?

    Ans. Latching currents is the minimum current which keeps the device latched during turn on. Holding current is the minimum current required to keep the device on.

24. Which device is suitable for high frequency applications?

    Ans. MOSFET (100 kHz to 1 MHz)

2

Triggering Circuits

The circuit which applies positive voltage to the gate of SCR is called triggering circuit. The basic triggering circuits used for turning the SCR on are:

(1) R-Triggering circuit
(2) RC-Triggering circuit
(3) UJT-Triggering circuit

Microprocessor based triggering circuits are discussed in Chapter 12.

2.1 R-TRIGGERING

The circuit of half controlled rectifier with R-triggering is shown in Fig. 2.1.

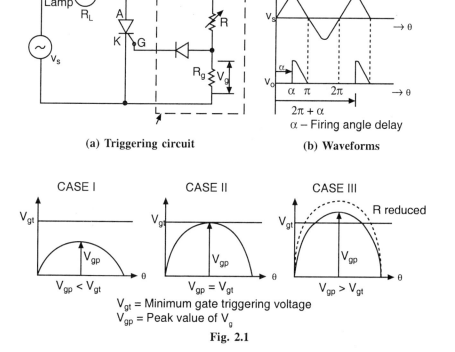

(a) Triggering circuit

(b) Waveforms

α – Firing angle delay

V_{gt} = Minimum gate triggering voltage
V_{gp} = Peak value of V_g

Fig. 2.1

During the positive half cycle, the SCR is triggered at α. It conducts from α to π. During the negative half cycle, it does not conduct since it is reverse biased. In the next positive half cycle, the SCR is triggered at $(2\pi + \alpha)$. It conducts from $(2\pi + \alpha)$ to 3π. The voltage across the load is positive segments of supply voltage. From the circuit shown in Fig. 2.1a.

$$I = \frac{V_s}{R_L + R + R_g} \qquad (2.1)$$

$$V_g = I \cdot R_g$$

V_{gt} = minimum gate triggering voltage and V_{gp} = Peak value of V_g.

The current through R_g can be varied by varying the resistance R. If the resistance is large, the current and V_g is less. In case (1), $V_{gp} < V_{gt}$, Therefore the SCR cannot be turned "on". If the resistance "R" is moderate, V_g is such that $V_{gp} = V_{gt}$. Therefore the SCR turns on at $\alpha = 90°$ as given in case II. If the resistance R is further reduced V_g increases. Therefore the SCR gets triggered at an angle α. This is given in case III. The value of α can be varied by varying the resistance R. The firing angle variation is from 0 to 90°. The diode will ensure that only positive voltage is applied at the gate. During the negative half cycle the diode blocks the negative voltage.

2.2 R-C TRIGGERING

R-C triggering circuit is shown in Fig. 2.2a. Consider the vector diagram of R-C circuit shown in Fig. 2.2(b). From this figure, it can be seen that capacitor voltage lags the supply voltage by an angle θ_1. The θ_1 can be varied by varying the resistance R. During the positive half cycle, current flows through R and C. At $\theta = \alpha$, v_c reaches V_{gt}. SCR turns on. It conducts from α to π. At $\theta = \pi$, $i = 0$, SCR turns off.

During the negative half cycle, the SCR is reverse biased. The capacitor voltage is such that the diode is reverse biased. Therefore SCR cannot be turned on during negative half cycle.

The firing angle delay is varied by varying R. In an RC circuit, the time delay is RC. Therefore the delay can be varied by varying R or C. With resistance alone, a variation from 0° to 90° can be obtained. With capacitance another 90° variation can be achieved. Therefore the total firing angle delay can be in the range of 0 to 180°.

2.3 UJT TRIGGERING

Unijunction transistor (UJT) consists of emitter, base 1 and base 2. The equivalent circuit has a diode and two resistors. The centre point of the resistors is called interbase point. If the emitter potential is less than the peak potential, UJT acts as a open switch. If the emitter potential is higher than the interbase potential, the UJT is on and the resistance value of R_{B1} is less. The ratio of voltage across R_{B1} to the supply voltage is defined as intrinsic stand off ratio.

When the battery in Fig. 2.3d is connected, the capacitor charges exponentially through the resistance R. The emitter of UJT is at capacitor voltage. The interbase

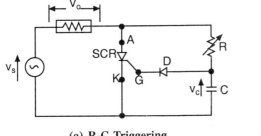
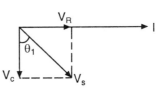

(a) R-C Triggering   (b) Vector diagram of R-C circuit

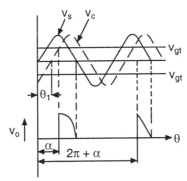

(c) Wave forms with R-load

Fig. 2.2

point is at a potential V_p (peak voltage). As long as the anode voltage is less than the peak potential, the diode is OFF. When the capacitor voltage goes more than V_p, the diode is ON. The capacitor quickly discharges through (R_1). The discharging is very fast since the resistance R is very less. When the capacitor voltage goes below V_v (valley Voltage), the diode is off since its anode potential is less than the cathode potential. The capacitor again gets charged and the above process repeats. When the capacitor is charging, there is no current through R_1. It has a discharging current only during the ON period. Thus the voltage across R_1 is a train of pulses which can be used for triggering the SCR. The firing angle delay is varied by varying the value of the resistance R. This circuit is called relaxation oscillator circuit since UJT relaxes most of the time.

$$\eta = \frac{V_{RB1}}{V_{BB}} = \frac{IR_{B1}}{I(R_{B1} + R_{B2})}; \quad \eta = \frac{R_{B1}}{R_{B1} + R_{B2}} \qquad (2.2)$$

Value of η varies from 0.6 to 0.7

In R-C circuit, capacitor voltage follows the equation

In Fig. 2.3f, $\qquad v_c = V_{BB}(1 - e^{-t/RC})$

When $t = T_c$, $v_c = V_P$

$$V_P = V_{BB}(1 - e^{-T_c/RC}); \quad \frac{V_P}{V_{BB}} = (1 - e^{-T_c/RC})$$

Substitute $\dfrac{V_P}{V_{BB}} = \eta; \quad T \simeq T_c$

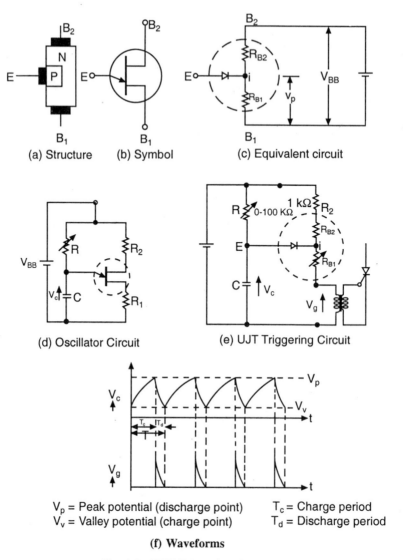

V_p = Peak potential (discharge point) T_c = Charge period
V_v = Valley potential (charge point) T_d = Discharge period

(f) **Waveforms**

Fig. 2.3 UJT Triggering Circuit

Neglect discharging period T_d since it is very small, therefore $\eta = 1 - e^{-T/RC}$

$$e^{-T/RC} = 1 - \eta$$

Taking log on both sides

$$-\frac{T}{RC} \ln e = \ln(1 - \eta); \quad \ln_e e = 1$$

$$-T = RC \ln(1 - \eta); \quad T = -RC \ln(1 - \eta)$$

$$T = RC \ln 1/(1 - \eta); \quad T = \text{Time period}; \quad \text{Frequency } (f) = 1/T$$

Problems

Ex. 2.1 Estimate the minimum and maximum charging resistor of UJT circuit for control of α between 20° to 160° of 50 Hz supply. Assume $C = 0.1 \ \mu F$ and $\eta = 0.7$.

Solution

$$T = \frac{1}{f} = \frac{1}{50} = 20 \text{ msec}$$

10 msec → 180° (∗ 20 msec for 360°)

$T_1 \to 20°$ $\qquad \frac{T_1}{10} = 20/180$

Therefore $T_1 = 1.11 \ \mu\text{sec}$

$$\frac{T_2}{10} = \frac{160}{180}; \qquad T_2 = 8.88 \text{ msec}; \qquad T_1 = R_1 C \ln \frac{1}{1-\eta}$$

$$1.11 * 10^{-3} = R_1 * 0.1 * 10^{-6} \frac{1}{10(1-0.7)}; \qquad R_1 = 9.25 \ k\Omega$$

$$T_2 = R_2 C \ln \frac{1}{1-\eta}; \qquad 8.88 * 10^{-3} = R_2 * 0.1 * 10^{-6} \ln \frac{1}{1-0.7}$$

$$R_2 = 74 \ K\Omega$$

Resistance has to be varied from 9.25 KΩ to 74 KΩ to vary α from 20° to 160°.

Short Questions and Answers

1. What is the range of firing angle in the case of: (i) R-Triggering, (ii) R-C Triggering and (iii) UJT Triggering.
   Ans. (i) 0-90°, (ii) 0-180° (iii) 0-180°
2. What is the function of diode in R-C triggering circuit?
   Ans. Diode ensures that the SCR is turned on only in positive half cycle.
3. What is the advantage of UJT triggering circuit over R-C triggering circuit?
   Ans. UJT triggering circuit provides isolation between control circuit and power circuit using a pulse transformer. Isolation is not present in the case of R-C triggering circuit.
3. Why R-triggering circuit is not popular?
   Ans. The maximum value of firing angle is 90° only. It is not popular since phase control is possible only in the range 0 to 90°.

3

Commutation Circuits

3.1 INTRODUCTION

The gate has no control over the thyristor once the thyristor starts conducting. The flow of current through the thyristor can be stopped by some external method. The turn off process of a thyristor is called commutation.

The generation of triggering pulses involves simple low power circuits. But the commutation circuits which handle large power are complicated. It may be noted that it is more challenging to turn off a thyristor than to turn it on. Most difficulties in the design of power electronic circuits and a large number of their faults may be due to the forced commutation circuits.

3.2 COMMUTATION

It is necessary to apply a reverse voltage to turn off a thyristor in minimum time. When the anode is made negative with respect to cathode, the holes around J_1 travel towards negative terminal and the electrons around J_3 travel towards positive terminal of the battery as shown in Fig. 3.1a. This results in a small reverse current in the external circuit. After the holes and electrons around J_1 and J_3 have been removed, the reverse current stops and J_1 and J_3 assume blocking state.

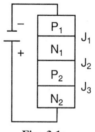

Fig. 3.1a

The fundamental concepts of series L.C. circuit and ringing circuit are required for understanding various commutation circuits. They are discussed in the following sections.

3.3 SERIES LC CIRCUIT

If a capacitor is charged through an inductor and a unidirectional device it gets charged to twice the supply voltage by the end of the half cycle. All the elements

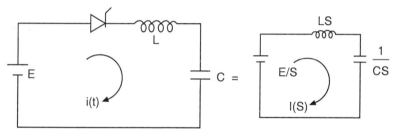

Fig. 3.1b Series LC Circuit

are assumed to be ideal and initial conditions are zero. Consider the circuit shown in Fig. 3.1b. The current in S-domain can be obtained using Ohm's law.

$$I(s) = \frac{E/S}{LS + 1/CS} = \frac{E/S}{(LCS^2 + 1)/CS} = \frac{EC}{LCS^2 + 1}$$

$$= \frac{EC}{LC\left(S^2 + \frac{1}{LC}\right)} = \frac{E}{L\left(S^2 + \frac{1}{LC}\right)}$$

Dividing and multiplying with $\sqrt{LC}$, we get

$$= \frac{E}{L}\frac{1}{\left(S^2 + \frac{1}{LC}\right)} * \frac{\sqrt{LC}}{\sqrt{LC}} = \frac{E}{L}\sqrt{LC}\frac{1/\sqrt{LC}}{\left(S^2 + \frac{1}{LC}\right)}$$

Substitute natural frequency $= \dfrac{1}{\sqrt{LC}} = \omega$

$$= E\sqrt{C/L}\,\frac{\omega}{S^2 + \omega^2}$$

Taking inverse Laplace transform $i(t) = E\sqrt{C/L}\sin \omega t$.

Capacitor Voltage $\quad v_c = \dfrac{1}{C}\displaystyle\int i\,dt$

$$v_c = \frac{1}{C}\int E\sqrt{C/L}\sin \omega t \cdot dt = \frac{E}{\sqrt{LC}}\int_0^t \sin \omega t \cdot dt$$

$$= E\omega \left.\frac{\cos \omega t}{\omega}\right|_0^t = E(1 - \cos \omega t) \qquad (3.1)$$

Substitute $\omega t = \theta$

Therefore $\quad v_c = E(1 - \cos\theta)$

When $\quad \theta = \pi$

Therefore $\quad V_c = E(1 - \cos \pi) = 2E$

$\quad V_c = 2E$

32 *Fundamentals of Power Electronics*

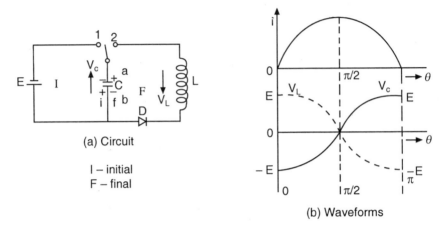

(a) Circuit

I – initial
F – final

(b) Waveforms

Fig. 3.2 Ringing circuit

The energy stored in the inductor transfers to the capacitor after the SCR is turned off. Therefore the capacitor voltage increases to 2E.

3.4 RINGING CIRCUIT

The ringing circuit is shown in Fig. 3.2. It is needed for understanding most of the commutation circuits.

In the ringing circuit, the capacitor is connected across the battery by closing the switch S to the position 1. Afterwards the switch is closed towards the positon 2. The circuit formed by precharged capacitor, inductor and the diode is called ringing circuit. In the figure 3.2a, loop F represents the ringing circuit. The initial polarity of the capacitor is such that the diode is forward biased. The current flows from the plate b, diode, L and C. We know that the current follows sine law. By the end of half cycle, the current reduces to zero and the capacitor gets charged in the opposite direction. In the closed loop $V_L + V_c = 0$. The wave form of V_L can be obtained using the equation $L\, di/dt$.

From 0 to $\pi/2$, di/dt is positive. Therefore V_L is positive. At $\pi/2$, di/dt is zero. $V_L = 0$. The wave form of V_c can be obtained using the equation $V_c = - V_L$. There cannot be a current in the negative direction since the diode blocks the reverse current.

3.5 TURN OFF METHODS

The current through the thyristor may be interrupted by means of a switch, either in series with the thyristor or in parallel with it. The series switch has to be opened for commutation and the parallel switch has to be closed for commutation. These methods are not generally used. The methods used are natural or line commutation and forced commutation. In ac circuits, the current through the thyristor has a natural zero crossing and the thyristor is reversed biased when the ac source reverses its polarity. This is called line commutation.

In dc applications, when the thyristor conducts, it continues to conduct until the current is interrupted or a reverse current is driven forcibly. The process of driving a reverse current and turning off a thyristor is called forced commutation. This can be classified into five different types:

Class A: Self commutated by resonating the load
Class B: Self commutated by an *LC* circuit
Class C: *C* or *LC* switched by another load carrying thyristor
Class D: *C* or *LC* switched by an auxiliary thyristor
Class E: External pulse commutation

3.5.1 Class A Commutation

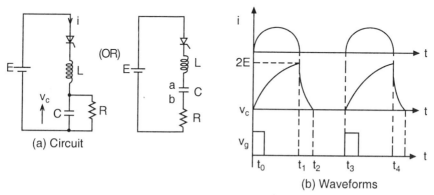

(b) Waveforms

Fig. 3.3   Class A Commutation

In this type, the load resistance can be connected either in series or in parallel with the commutation capacitor. The forced commutation circuit is shown in Fig. 3.3(a). The corresponding waveforms are shown in Fig. 3.3 (b). The elements *L* and *C* are designed such that the over all circuit is under damped and oscillations can be produced.

When the thyristor is triggered, the current flows and capacitor charges with plate a +ve. Because of the resonating action of *L* and *C*, the current reduces to zero and tries to flow in reverse direction. Hence the thyristor turns off. The capacitor charges nearly to 2*E* since it is charged through an inductor and a unidirectional device. The anode is at *E* with respect to ground while the cathode is at 2*E* with respect to the ground. Hence the thyristor is reverse biased.

In Fig 3.3b, the current follows sine law since the thyristor, *L* and *C* form a ringing circuit. At t_1, the current reaches zero. The capacitor voltage reaches 2*E* since V_c follows cosine law. This process repeats if the thyristor is triggered by applying positive pulse between gate and cathode at t_3. Thus it is a self commutating circuit since the thyristor turns off automatically after it has been turned on.

3.5.2 Class B Commutation

A *LC* circuit is connected in parallel with the thyristor as shown in Fig. 3.4(a). The corresponding wave forms are shown in Fig. 3.4(b). Before the thyristor is turned on, the capacitor charges with plate *a* +. At $t = t_0$, when the thyristor

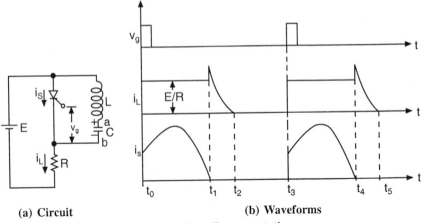

(a) Circuit (b) Waveforms

Fig. 3.4 Class B commutation

is triggered, load current flows and also L and C form a closed circuit. The thyristor current is the sum of load current and capacitor current during $t_0 - t_1$. Therefore the thyristor current increases with the increase in the capacitor current. The thyristor current can be obtained by adding ringing current with the constant load current. The capacitor discharges and by ringing its charge reverses and when the current tries to reverse, the thyristor turns off. The resonant current must be greater than the load current for successful commutation. The final polarity of the capacitor is such that it reverse biases the thyristor. The reverse bias decreases as the capacitor discharges. The capacitor voltage reduces to zero and it again charges with b +. The circuit turn off time t_c should be greater than the device turn off time for successful commutation.

In class B also, the thyristor conducts for a finite period and then it automatically turns off. It may be noted that the commutating components do not carry the load current.

3.5.3 Class C Commutation

The class C commutation circuit and the waveforms are shown in Fig. 3.5(a) and 3.5(b) respectively. The loads R_1 and R_2 are alternately fed from S_1 and S_2 respectively. At $t = t_0$, thyristor 1 is triggered. The capacitor is charged with plate b + through R_2. If the thyristor 2 is turned on at t_1, the capacitor voltage reverse biases S_1 through the conducting thyristor S_2. The capacitor discharges through S_1 and S_1 is turned off. The capacitor maintains reverse bias after S_1 is turned off. Thus turning on of one thyristor causes the commutation of other thyristor. When S_2 is conducting, the capacitor charges to a + through R_1.

3.5.4 Class D Commutation

The class D commutation circuit and the corresponding waveforms are shown in Fig. 3.6(a) and 3.6(b) respectively. S_m and S_A are the main and auxiliary thyristors respectively. The capacitor is initially charged with plate a +ve by triggering S_A alone. S_A turns off naturally when the capacitor has fully charged and the capacitor current falls below the holding current of thyristor.

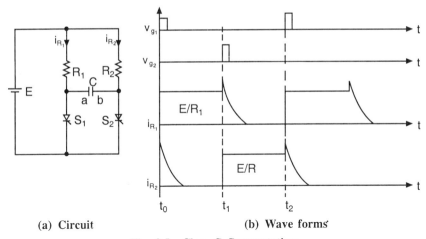

(a) Circuit (b) Wave forms

Fig. 3.5 Class C Commutation

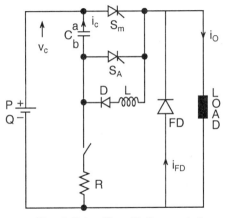

Fig. 3.6(a) Class D Commutation

The thyristor S_1 is triggered at $t = 0$, the load current flows through E, S_m and R. The elements S_m, L, D and C form a ringing circuit. The commutation current flows through S_1, L, D and C. In the duration, $0–t_1$, the thyristor current is the sum of load current and ringing current. The polarity of the capacitor voltage is reversed due to ringing. This polarity can not reverse further due to the presence of D. Thus capacitor current flows from 0 to t_1 only. The new polarity of capacitor ($b+$) forward biases the thyristor S_A. S_A is triggered at t_2. Now the capacitor drives a current through the path b-S_A-S_m-a. A large current flows through S_m in the opposite direction since the current is limited only by the forward resistance of the devices. Hence S_1 turns off by the principle of voltage commutation. Now the capacitor charges to $a +$ since the current enters the terminal a. Complete explanation of this circuit is given in section 5.5 of Chapter 5.

3.5.5 Class E Commutation

It is also called external pulse commutation. The circuit and waveforms are

36 *Fundamentals of Power Electronics*

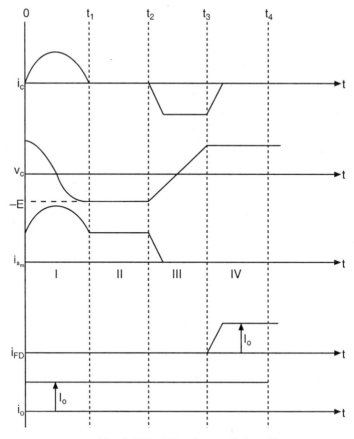

Fig. 3.6(b) Waveforms of class D

shown in Fig. 3.7. An external pulse source produce the energy required for commutation. The source supplies a pulse to commutate the thyristor. When the dot end of winding II is positive, the dot end of winding I is also positive.

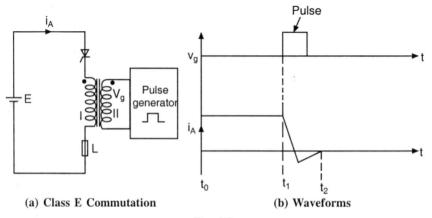

(a) Class E Commutation   (b) Waveforms

Fig. 3.7

The winding I pumps a current through the SCR and it gets turned off. The voltage across winding I reverse biases the SCR.

3.5.6 Class F Commutation

The circuit and waveforms are shown in Fig. 3.8. Class F commutation is called line commutation or natural commutation. At $\theta = \alpha$, the SCR is triggered. The SCR conducts from α to π. At $\theta = \pi$, the current reduces to zero. The SCR naturally gets turned OFF. From π to 2π the source applies reverse bias to the SCR. This commutation is called line commutation since the line voltage applies reverse bias to the SCR.

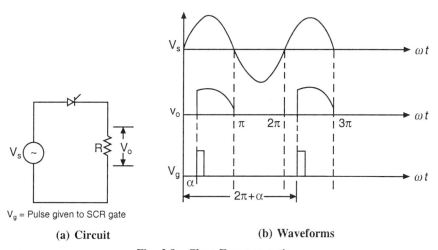

V_g = Pulse given to SCR gate

(a) Circuit (b) Waveforms

Fig. 3.8 Class F commutation

Short Questions and Answers

1. What is the peak value of SCR current in class B commutation?

   Ans. $I_P = \dfrac{E}{R} + E\sqrt{\dfrac{C}{L}}$

2. Compare voltage commutation with current comutation.

   Ans. In voltage commutation, a precharged capacitor drives a large current through the conducting SCR and turns it off. It is a fast method of commutation.

   In current commutation, precharged capacitor in series with an inductor drives current through the conducting SCR and turns it off. It is shower than voltage commutation.

3. What is the most popular method of forced commutation?

   Ans. Class D.

4. Why class F commutation is called line commutation?

   Ans. Line voltage applies reverse bias to the SCR.

5. Why class A commutation is called load commutation?

   Ans. The load resistance, L and C form under damped circuit and produces current zero.

4

Phase Controlled Rectifiers

4.1 INTRODUCTION

A rectifier is a power controller that converts ac voltage into dc voltage. They are used in the following areas.

   (i) Speed control of dc motors
   (ii) Battery charger circuits
   (iii) Uninterruptible power supplies
   (iv) Electroplating
   (v) Electrolysis

4.2 CLASSIFICATIONS OF RECTIFIERS

Based on the characteristics, the rectifier circuits are classified as follows:

(i) Based on the controllability of output voltage

In some rectifier circuits, the output voltage cannot be controlled and they are called as uncontrolled rectifiers. They use only diodes in the circuit. In some rectifiers, the output voltage is controllable and they are called as controlled rectifiers. They use either thyristors alone or combination of thyristors and diodes.

(ii) Based on the rectification of input waveform

If a circuit produces dc output voltage corresponding to only one half cycle of input voltage, it is called as half wave rectifier. If it produces output voltage corresponding to both positive and negative half cycles, it is called as full wave rectifier.

(iii) Based on the number of phases of input voltage

The ac voltage applied to the rectifier may be of single phase, three phase or in general polyphase. Based on the number of phases, the rectifiers are called single phase or polyphase rectifiers.

(iv) Based on the number of repetitive segments in output voltage waveform

The output voltage waveform consists of segments of input ac voltage and these segments repeat over one cycle of input voltage. Depending upon the number of segments, the rectifier is classified as single pulse, two pulse, three pulse, six pulse or twelve pulse rectifiers.

(v) Based on volt-ampere characteristic

The current on the dc side of a rectifier remains in the same direction due to the unidirectional devices such as diodes and thyristors used in rectifier circuit. If the polarity of the average output voltage also remains unchanged, the volt-ampere characteristic is confined to only one quadrant and the rectifier is called as single quadrant rectifier. If the voltage polarity reverses, it operates in two quadrants and the rectifier is called as two quadrant rectifier. If two such converters are connected in antiparallel, both the voltage and current can be reversed. In this case the converter is called as four quadrant converter or dual converter. The single quadrant rectifier uses either only diodes or diodes and thyristors. It is called as semi controlled converter. The two quadrant rectifier uses only SCRs and it is called as fully controlled converter.

(vi) Based on the direction of current in ac lines

During the operation of the rectifier, if current in all flows in only direction, it is called as single way rectifier. On the other hand, if current flows in both the directions, it is called as two way rectifier.

4.3 PERFORMANCE OF RECTIFIERS

If the output voltage of a rectifier is of constant magnitude as in a battery and it draws sinusoidal current at unity power factor from the ac source, then the rectifier is said to have the best performance. But it is not so in practical rectifiers. The output voltage of a practical rectifier has segments of the input ac voltage and hence it contains both dc voltage and ac voltage. The current drawn from the ac source is not sinusoidal. It contains fundamental component and harmonics. The power factor at the input of the rectifier is poor even for resistive load on dc side and also the power factor decreases as the triggering angle is delayed. The amount of these ac voltage, harmonic current, input power factor etc. decides the quality of the rectifiers. They are calculated based on certain quantities measured on ac and dc sides.

A. Quantities on output side

(i) dc quantities on output side

The average dc output voltage = V_{dc}
The average dc output current = I_{dc}
The output dc power = $P_{dc} = V_{dc}I_{dc}$

(ii) Total rms quantities on output side

Since the output contains both dc and ac components, the total rms values are symbolised as follows

The rms output voltage = V_{rms}
The rms output current = I_{rms}
The output ac power = $P_{ac} = V_{rms}I_{rms}$

(iii) Efficiency of rectification

The power expected from a rectifier is due to the average voltage and the average current, that is, P_{dc} whereas it produces a total power P_{ac} due to both dc and ac components at the output. The power due to ac components cannot be utilised in applications such as battery charger, dc motor speed control, electroplating, electromagnetics etc. The efficiency of the rectifier is decided based on P_{dc} and P_{ac}. The efficiency of rectification or rectification ratio η is defines as

$$\eta = \frac{P_{dc}}{P_{ac}} = \frac{V_{dc} I_{dc}}{V_{rms} I_{rms}} \qquad (4.1)$$

(iv) AC components on output side

The rms value of the output voltage is due to both dc and ac components. The rms value of ac component alone is obtained by the expression

$$V_{ac} = \sqrt{(V_{rms}^2 - V_{dc}^2)}$$

A good rectifier produces as low ac voltage (V_{ac}) as possible. This increases the efficiency of the rectifier and also makes filtering easy.

(v) Form Factor

The quality of the output voltage depends on the relative values of dc and ac component and is defined by form factor. The form factor is defined as

Form Factor (FF) = V_{rms}/V_{dc}

For a good rectifier, the form factor is around unity.

(vi) Ripple Factor

The rectified output voltage varies with time due to ac component. The difference between its maximum and minimum values is called as the peak-to-peak ripple voltage. The ripple factor is defined as

$$\text{Ripple Factor (RF)} = V_{ac}/V_{dc} = \sqrt{\left(\frac{V_{rms}}{V_{dc}}\right)^2 - 1} \qquad (4.2)$$

B. Quantities on input side

(i) Displacement power factor

The voltage applied to a rectifier is sinusoidal whereas the current drawn from the ac source is non sinusoidal due to the switching action of the rectifying elements. This non sinusoidal current has a fundamental component. The phase angle difference between the fundamental component and the input voltage is called as displacement angle. In the case of controlled rectifier, the displacement angle increases with the increase in triggering angle delay. The displacement power factor or simply displacement factor (DF) is defined as

$$DF = \cos \phi$$

where ϕ is the angle between the zero crossing of the input ac voltage and the fundamental component of input current.

Phase Controlled Rectifiers

(ii) Harmonic current in ac source

Due to non sinusoidal nature of the input current, it has fundamental component and harmonics. The harmonic current I_h is obtained by subtracting the fundamental component I_1 from the total rms input current I_3. The amount of harmonic current is defined by harmonic factor (HF) which can be calculated from the expression as

$$\text{HF} = I_h/I_1 = \sqrt{\left(\frac{I_3}{I_1}\right)^2 - 1} \qquad (4.3)$$

The lower the value of the harmonic factor, the better is the quality of the rectifier.

(iii) Input power factor

In general the power factor is defines as

$$\text{Power factor} = \frac{\text{Real power}}{\text{Apparent power}}$$

The sinusoidal input voltage and the fundamental component of current alone produce real power. The apparent power is the product of the rms value of voltage V_3 and the rms value of current I_3 which includes the harmonic current also. The power associated with the harmonic current with fundamental voltage (sinusoidal voltage) does not constitute any real power. Therefore

$$\text{Real power} = V_3 I_1 \cos\phi$$

Apparent power = $V_3 I_3$

$$\text{Input power factor} = \frac{V_3 I_1 \cos\phi}{V_3 I_3} = \frac{I_1}{I_3} \cos\phi \qquad (4.4)$$

(iv) Transformer utilization factor

Current flows through the windings of a transformer for full cycle when the transformer supplies normal loads. When a transformer supplies a rectifier load, current flows only for part of cycle and hence the windings are not fully utilized. The extent to which the transformer is utilized is given by the term transformer utilization factor (TUF) which is expressed as

$$\text{TUF} = \frac{P_{dc}}{m V_3 I_3} \qquad (4.5)$$

where V_3 is the rms value of transformer phase voltage
I_3 is the total rms current in the transformer phase winding
m is the number of phases of transformer.

It is to be noted that the unit of P_{dc} is in watts and that of ($m\, V_3 I_3$) is in volt-amperes.

4.4 SINGLE PHASE RECTIFIERS

A single phase ac supply feeds power to these rectifiers. These rectifiers are classified depending upon certain characteristics. Depending on the rectification on the half cycles of input ac voltage, they are called as half wave rectifier and full wave rectifier. The supply to the rectifier can be through a centre tapped transformer, in which case it is called as centre tapped transformer rectifier. If the rectifying devices are connected in the form of a bridge, it is called as single phase bridge rectifier.

Single phase uncontrolled rectifier uses only diodes where as single phase controlled rectifier use thyristors with or without diodes. If only thyristors are used it is called as fully controlled rectifier and if a combination of thyristors and diodes are used it is called as semi controlled or half controlled rectifier.

4.5 HALF CONTROLLED RECTIFIER WITH R-LOAD

It is called half controlled rectifier since the thyristor conducts only for positive half cycle. At $\theta = \alpha$, the SCR is triggered and it conducts up to π. SCR does not conduct from π to 2π since it is reverse biased. If the firing angle delay is large, the average output voltage gets reduced since the conduction period is reduced. The starting point of the current waveform can be controlled by controlling the firing angle. Hence this type of rectifiers are called phase controlled rectifiers. The circuit is shown in Fig. 4.1.

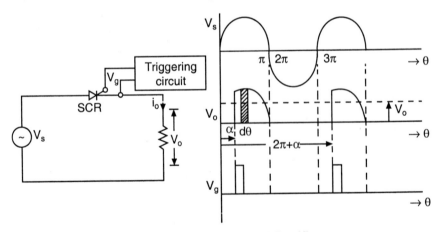

Fig. 4.1(a)   Half Controlled Rectifier

The expression for the average voltage is as follows:

The average value of a periodic waveform can be obtained by dividing the area with the base. Base is the width corresponding to one cycle.

$$V_0 = \frac{\text{Area}}{\text{Base}} = \frac{1}{2\pi} \int_\alpha^\pi v_s \, d\theta = \frac{1}{2\pi} \int_\alpha^\pi V_m \sin \theta \, d\theta = \frac{V_m}{2\pi} |\cos \theta|_\alpha^\pi$$

$$= \frac{V_m}{2\pi} (\cos \alpha - \cos \pi) = \frac{V_m}{2\pi} (1 + \cos \alpha) \qquad (4.6)$$

4.6 HALF CONTROLLED RECTIFIER WITH R-L LOAD

The current through SCR starts flowing form α since it is triggered at α. When the SCR is conducting, voltage across the load is same as the source voltage. With RL load, the SCR conducts beyond 180° since the inductor voltage forward biases the SCR and keeps it on. From α to π, V_0 and i are positive. Power is transferred from source to the load. From π to β, i is positive and V_0 is negative. The power is fed back to the source. This is called recycling. Recycling continues till the energy in the inductor is given back to the source. The average output voltage gets reduced due to the negative segments. The larger the inductance, the larger the negative area and vice versa. The waveforms are shown in Fig. 4.2b.

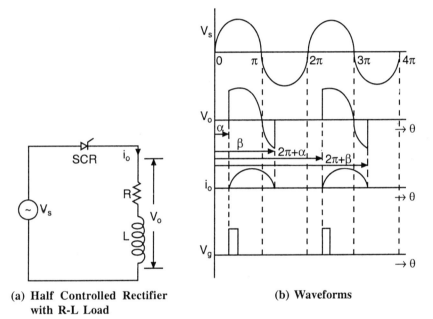

(a) Half Controlled Rectifier with R-L Load

(b) Waveforms

Fig. 4.2

$$V_0 = \frac{1}{2\pi} \int_\alpha^\beta v \, d\theta = \frac{1}{2\pi} \int_\alpha^\beta V_m \sin \theta \, d\theta$$

$$V_0 = \frac{V_m}{2\pi} |\cos \theta|_\beta^\alpha = \frac{V_m}{2\pi} (\cos \alpha - \cos \beta)$$

$$V_0 = \frac{V_m}{2\pi} (\cos \alpha - \cos \beta) \tag{4.7}$$

4.7 1-ϕ FULLY CONTROLLED RECTIFIER WITH R-LOAD

During the positive half cycle, the SCR1 is forward biased and it conducts from α to π. The current flows from the dot end of the section A through S_1, R to the non dot end. The direction of output voltage is from left to right.

In the negative half cycle, the SCR2 is forward biased. It conducts from

$\pi + \alpha$ to 2π since it is triggered at $\pi + \alpha$. The current flows in the path-non dot end of section B, $S2$, R, dot end. Once again the direction of load current and the direction of output voltage remain the same as previous half cycle. Thus the rectifier converts bidirectional current into unidirectional current. The circuit is shown in Fig. 4.3.

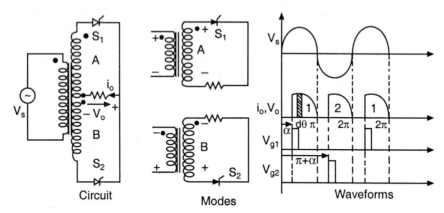

Fig. 4.3 Single phase fully controlled rectifier with R-load

It is called fully controlled rectifier since the output can be controlled during the full cycle of the supply. It is also called a two pulse converter. The average output voltage of this circuit is twice that of the half controlled rectifier with R load.

$$V_0 = \frac{1}{\pi} \int_\alpha^\pi v\, d\theta = \frac{1}{\pi} \int_\alpha^\pi V_m \sin\theta\, d\theta = \frac{V_m}{\pi} \left| -\cos\theta \right|_\alpha^\pi$$

$$= \frac{V_m}{\pi} (\cos\alpha - \cos\pi) = \frac{V_m}{\pi} [\cos\alpha - (-1)]$$

$$V_0 = \frac{V_m}{\pi} (1 + \cos\alpha) \tag{4.8}$$

4.8 1-ϕ FULLY CONTROLLED RECTIFIER (BRIDGE TYPE)

Bridge type rectifier shown in Fig. 4.4 has two legs. Each leg consists of two SCRs. Number them such that odd numbers are at the top and even numbers are at the bottom. Sum of the numbers in each leg is equal to 5. During the positive half cycle, the SCRs S_1 and S_2 are forward biased. If they are simultaneously triggered, the current flows through the path Ph-S_1-R-S_2-N. The voltage across the load is a segment of the supply voltage. During the negative half cycle the SCRs S_3 and S_4 are forward biased. When they are simultaneously triggered, the current flows through the path NS_3-S_4-Ph. The voltage across the load is a segment of the supply voltage.

The PIV (peak inverse voltage) in the case of bridge rectifier is V_m (PIV is the largest voltage that can appear across the non conducting SCR). In the case of centre tapped type, the PIV is $2V_m$.

The expression for average output voltage is $(V_m/\pi)(1 + \cos\alpha)$.

Phase Controlled Rectifiers 45

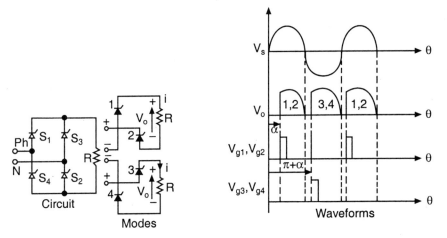

Fig. 4.4 Bridge type fully controlled rectifier

4.9 1-φ FULLY CONTROLLED RECTIFIER WITH R-L LOAD

The waveforms of bridge type and centre tapped type fully controlled rectifier are similar. Therefore they are combined and shown in Fig. 4.5a.

At α, the SCRs S_1 and S_2 are triggered. They conduct from α to $(\pi + \alpha)$. At $(\pi + \alpha)$, S_3 and S_4 are triggered. The incoming thyristors reverse bias the out going thyristors and S_1 and S_2 get turned off by line commutation. The thyristors S_3 and S_4 conduct from $(\pi + \alpha)$ to $(2\pi + \alpha)$ and the above process repeats. The waveforms for output voltage and current are drawn by assuming a large inductance and negligible resistance. The circuit and waveforms are shown in Fig. 4.5(a).

In the case of centre tapped type, the SCRs T_1 and T_2 conducts alternately and the waveforms are similar to that of bridge type.

$$V_0 = \frac{1}{\pi} \int_\alpha^{\pi+\alpha} v \, d\theta = \frac{1}{\pi} \int_\alpha^{\pi+\alpha} V_m \sin \theta \, d\theta$$

$T_1, T_2 \Rightarrow$ for centre tapped rectifier; 1,2,3,4 $\Rightarrow$ for bridge rectifier

Fig. 4.5(a) Fully controlled rectifier with R-L load

$$V_0 = \frac{V_m}{\pi}\left|-\cos\theta\right|_\alpha^{\pi+\alpha} = \frac{V_m}{\pi}\left|\cos\theta\right|_{\pi+\alpha}^{\alpha}$$

$$= \frac{V_m}{\pi}[\cos\alpha - \cos(\pi + \alpha)] = \frac{V_m}{\pi}[\cos\alpha - (-\cos\alpha)]$$

$$V_0 = \frac{2V_m}{\pi}\cos\alpha \tag{4.9}$$

When α is between 0 and 90°, the converter acts as a rectifier. Power $P_0 = V_0 I_0$ where V_0, I_0 and P_0 are all positive. The power flows from a.c. side to the d.c. side. If the converter is operating with α between $\frac{\pi}{2}$ to π, the average output voltage is negative. The converter operates as an inverter. The power flows from d.c. side to the a.c. side. This is called regeneration. When the converter is operating between 90° and 180°, it is called line commutated inverter. The examples for regeneration are an electric locomotive going over a down hill and a lift that is coming down. Regeneration of the power is possible provided the source is in a position to absorb the power.

The waveforms for discontinuous conduction are shown in Fig. 4.5(b). At α, S_1 and S_2 are triggered and they conduct up to β. At β, the current reduces to zero. At $\pi + \alpha$, S_3 and S_4 are triggered and the above process repeats. From the current waveform it can be that current is discontinuous.

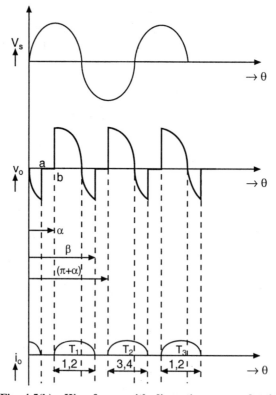

Fig. 4.5(b) Waveforms with discontinuous conduction

4.10 FULLY CONTROLLED RECTIFIER WITH SOURCE INDUCTANCE

The operation of the full converter can be understood by using the equivalent circuit shown in Fig. 4.6b. The source V_1 forward biases S_1, S_2. The source V_2 forward biases S_3, S_4. V_2 is in antiphase with V_1. The SCRs S_1 and S_2 conduct up to $\pi + \alpha$. At $\pi + \alpha$, S_3 and S_4 are triggered. The current cannot instantaneously transfer from S_1, S_2 to S_3, S_4 due to the presence of source inductance. The current through S_1 and S_2 gradually decreases and the current through S_3 and S_4 gradually increases such that the total current is equal to constant load current. The current is assumed to be constant since a highly inductive load

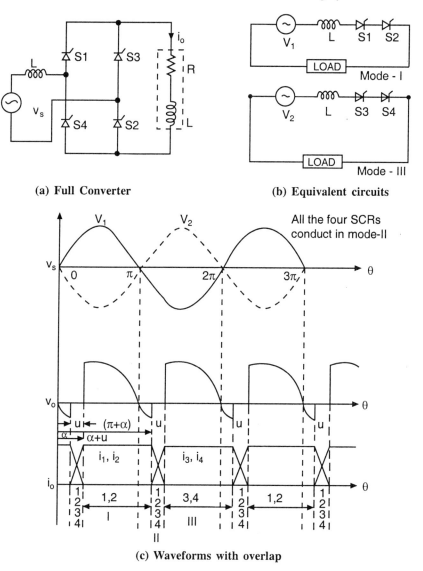

(a) Full Converter   (b) Equivalent circuits

(c) Waveforms with overlap

Fig. 4.6

is connected at the output (dc side). The period for which all the devices (S_1, S_2, S_3 and S_4) conduct is called overlap period. The corresponding angle is 'u'. From $\pi + \alpha + u$, the SCRs S_3 and S_4 conduct and the above process repeats.

$$V_{du} = \frac{1}{\pi} \int_{\alpha}^{\pi+\alpha} v \, d\theta$$

α to $(\pi + \alpha)$ = α to $(\alpha + u)$ + $(\alpha + u)$ to $(\pi + \alpha)$
It is sufficient if we integrate from $(\alpha + u)$ to $(\pi + \alpha)$, since the function does not exist from α to $(\alpha + u)$

$$V_{du} = \frac{1}{\pi} \int_{\alpha+u}^{\pi+\alpha} V_m \sin\theta \, d\theta$$

$$= \frac{V_m}{\pi} |-\cos\theta|_{\alpha+u}^{\pi+\alpha} = \frac{V_m}{\pi} [\cos(\alpha+u) - \cos(\pi+\alpha)]$$

$$V_{du} = \frac{V_m}{\pi} [\cos\alpha + \cos(\alpha+u)] \tag{4.10}$$

where V_{du} is the average output voltage by considered overlap.

The effects of overlap are (a) Average output voltage gets reduced. (b) The power factor gets reduced.

4.11a 1-φ FULLY CONTROLLED RECTIFIER WITH FREE WHEELING DIODE

At $\theta = \alpha$, S_1 and S_2 are triggered. They conduct from α to π. Beyond π, di/dt is negative. The voltage across the inductor forward biases the freewheeling diode. The diode D conducts until the energy in the inductance reduces to zero. At $(\pi + \alpha)$, S_3 and S_4 are triggered.

The circuits and waveform are shown in Fig. 4.7. with large inductive load. The current transfers from $S_1 S_2$ to D and D to $S_3 S_4$ etc. When the diode conducts, the voltage across the load is zero. The advantages with freewheeling diode are (a) Average output voltage is increases due to elimination of negative area. (b) Power factor is improved since the diode prevents the power flow from d.c. side to a.c side. The disadvantage is that regeneration is not possible due to the presence of freewheeling diode.

$$V_0 = \text{Average output voltage} = \frac{\text{Area}}{\text{Base}} = \frac{1}{\pi} \int v \, d\theta$$

$$= \frac{1}{\pi} \int_{\alpha}^{\pi} V_m \sin\theta \, d\theta = \frac{V_m}{\pi} \int_{\alpha}^{\pi} \sin\theta \, d\theta$$

$$= \frac{V_m}{\pi} |-\cos\theta|_{\alpha}^{\pi} = \frac{V_m}{\pi} (\cos\alpha - \cos\pi)$$

$$V_0 = \frac{V_m}{\pi} [1 + \cos\alpha] \tag{4.11}$$

4.11b SEMICONVERTER

A semiconverter shown in Fig. 4.8 is formed by two SCRs and two diodes. At

Phase Controlled Rectifiers 49

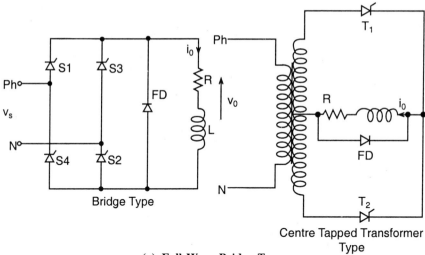

(a) Full Wave Bridge Type

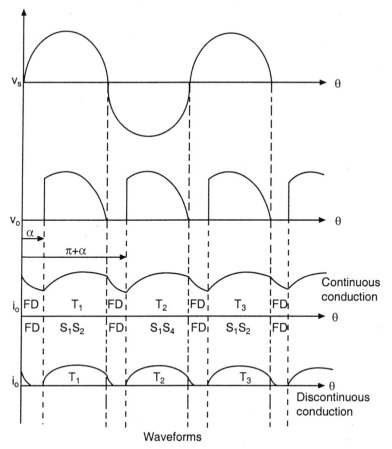

(b) Waveforms

Fig. 4.7

$\theta = \alpha$, S_1 is triggered. S_1 and D_2 conduct during the positive half cycle. S_1 continues to conduct beyond π since it is forward biased by the voltage across load inductor. From π to β, the diode D_1 is forward biased by the load voltage. Therefore the devices S_1 and D_1 conduct during this period. This is called inherent freewheeling. At $\theta = \pi + \alpha$, S_2 is triggered. S_2 and D_1 conducts from $\pi + \alpha$ to 2π. Later freewheeling takes place through S_2 and D_2. At $\pi + \alpha$, when S_1 is triggered the above process repeats.

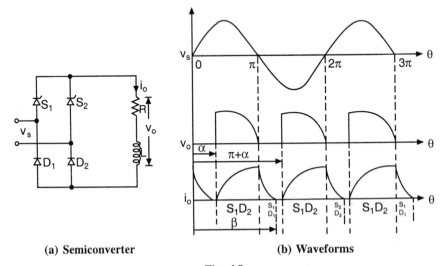

(a) Semiconverter (b) Waveforms

Fig. 4.8

In semi converter, the inherent freewheeling is not generally utilized. A separate freewheeling diode is connected across the load. This is because the inherent freewheeling increases the average current rating of the SCR.

$$V_0 = \text{Average output} = \frac{\text{Area}}{\text{Base}} = \frac{1}{\pi}\int_\alpha^\pi v\, d\theta = \frac{1}{\pi}\int_\alpha^\pi V_m \sin\theta\, d\theta$$

$$= \frac{V_m}{\pi}\int_\alpha^\pi \sin\theta\, d\theta = \frac{V_m}{\pi}[-\cos\theta]_\alpha^\pi$$

$$V_0 = \frac{V_m}{\pi}[1 + \cos\alpha] \qquad (4.12)$$

4.12 FULLY CONTROLLED RECTIFIER USING ONE SCR

The curcuit is shown in Fig. 4.9. At $\theta = \alpha$, the SCR-1 is triggered and the current flows through D_1, S_1, L and D_2. The voltage across the load will be a segment of the supply voltage. From π to $\pi + \alpha$, the free wheeling diode conducts. At $\pi + \alpha$, S_1 is triggered again. Now the current flows through D_3, S_1, L and D_4. Thus the output voltage is positive during the negative half cycle of the input voltage.

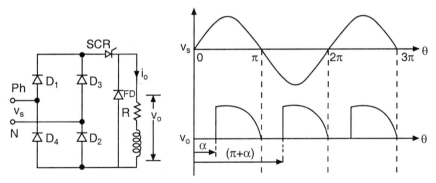

Fig. 4.9 Full converter with one SCR

4.13 3-φ HALF CONTROLLED CONVERTER (3-PULSE CONVERTER)

The output of this converter has 3-pulses/cycle. Hence it is called 3-pulse converter. At any time one SCR that has largest positive voltage at the anode conducts. At 30°, SCR-1 is triggered in the case of $\alpha = 0°$. At 150°, S_2 is triggered. From 150° to 180°, the value of v_b is more than v_a. Therefore the SCR T_1 is reverse biased and gets turned OFF. Thus triggering of one SCR turns OFF the outgoing SCR by the principle of line commutation. The point p is called natural point of commutation. In 3-φ converters, α is measured from the point p. The voltage will be positive segments of the a.c. input voltage. The circuit is shown in Fig. 4.10.

V_m = Peak value of phase voltage

$$V_0 = \frac{1}{2\pi/3} \int_{30+\alpha}^{150+\alpha} V_m \sin\theta \, d\theta = \frac{3V_m}{2\pi} [-\cos\theta]_{30+\alpha}^{150+\alpha}$$

$$= \frac{3V_m}{2\pi} [\cos\theta]_{150+\alpha}^{30+\alpha} = \frac{3V_m}{2\pi} [\cos(30+\alpha) - \cos(150+\alpha)]$$

$$= \frac{3V_m}{2\pi} [\cos 30 \cos\alpha - \sin 30 \sin\alpha - (\cos 150 \cos\alpha - \sin 150 \sin\alpha)]$$

$$= \frac{3V_m}{2\pi} [2\cos 30 \cos\alpha] = 3\sqrt{3} \frac{V_m}{2\pi} \cos\alpha \qquad (4.13)$$

4.14 3-φ FULLY CONTROLLED BRIDGE CONVERTER (6 PULSE CONVERTER)

This converter has three legs. Number the SCRs such that top ones are odd numbered and bottom ones are even numbered. Difference of the numbers in each leg is equal to 3. A 3-φ fully controlled converter has 6 thyristors as shown in Fig. 4.11(a). At any time two SCRs with one from the positive group and one from negative group conduct. The equivalent circuit for Fig. 4.11(a) is shown in Fig. 4.11(b). A 3-φ, 6 pulse converter can be understood by considering

52 *Fundamentals of Power Electronics*

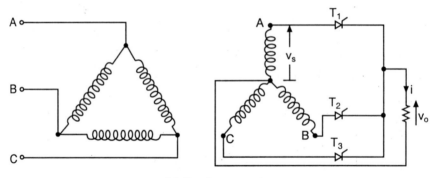

(a) 3-pulse converter

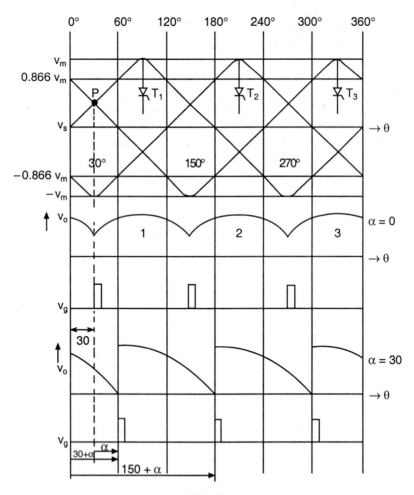

(b) Waveforms

Fig. 4.10

the circuit as two 3-pulse converters connected in series. Therefore the output of 6-pulse converter is twice as that of 3-pulse converter. Fig. 4.11(c) shows the equivalent circuit when SCRs 1 and 6 are conducting. It can be observed that the d.c. output voltage is a segment of a.c. line to line voltage.

The waveforms of the phase voltages are shown in Fig. 4.11(e). The thyristors are triggered in the sequence 56, 61, 12, 23, 34, 45, 56, 61. Pairs of SCRs are

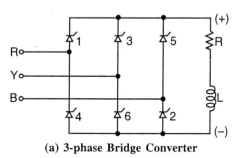

(a) 3-phase Bridge Converter

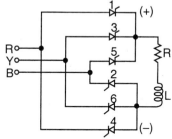

(b) 2 Nos. 3-pulse converter connected in Series as above is equivalent to 6 pulse converter

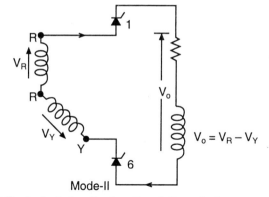

(c) Equivalent circuit when 1 and 6 conduct

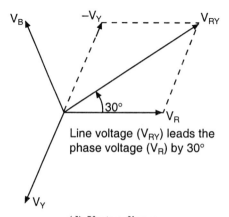

Line voltage (V_{RY}) leads the phase voltage (V_R) by 30°

(d) Vector diagram

54 *Fundamentals of Power Electronics*

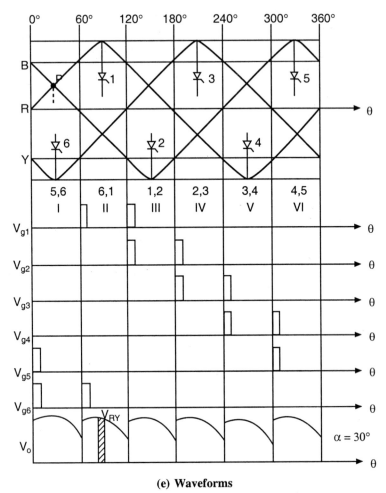

(e) Waveforms
Fig. 4.11

triggered so that a closed path is provided at starting. An SCR with maximum positive voltage at the anode of top group and maximum negative voltage at the cathode of bottom group conducts if they are triggered.

From vector diagram, the voltage across load is equal to

$$V_{RY} = \sqrt{3}\, V_m \sin(30 + \theta)$$

From Fig. 4.11(f) $V_0 = \dfrac{1}{\pi/3} \displaystyle\int_{30+\alpha}^{90+\alpha} \sqrt{3}\, V_m \sin(30 + \theta)\, d\theta$

$$= \dfrac{3\sqrt{3}\, V_m}{\pi} [-\cos(30 + \theta)]_{30+\alpha}^{90+\alpha}$$

$$= \dfrac{3\sqrt{3}\, V_m}{\pi} [\cos(\theta + 30)]_{90+\alpha}^{30+\alpha}$$

$$= \dfrac{3\sqrt{3}\, V_m}{\pi} [\cos(60 + \alpha) - \cos(120 + \alpha)]$$

Phase Controlled Rectifiers 55

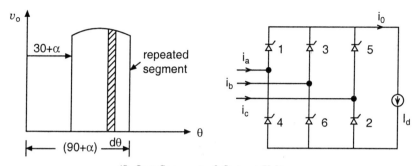

(f) One Segment of Output Voltage

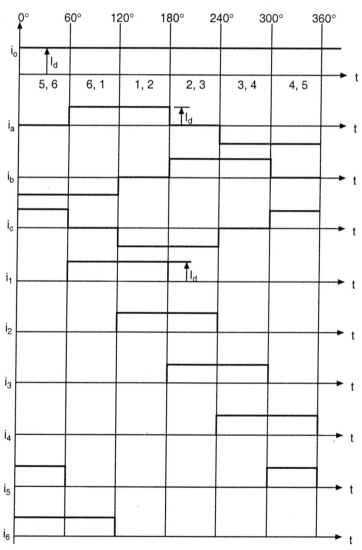

(g) Current Waveforms

$$V_0 = \frac{3\sqrt{3}\,V_m}{\pi} [\cos 60° \cos \alpha + \sin 60° \sin \alpha$$
$$- (\cos 120° \cos \alpha - \sin 120° \sin \alpha)]$$
$$= \frac{3\sqrt{3}\,V_m}{\pi} (\cos 60° \cos \alpha + \cos 60° \cos \alpha)$$
$$= \frac{3\sqrt{3}\,V_m}{\pi} (2 \cos 60° \cos \alpha) = \frac{3\sqrt{3}\,(V_m)}{\pi} \cos \alpha \qquad (4.14)$$

Where V_m is the peak value of the phase voltage.

$$V_0 = \frac{3\sqrt{3}}{\pi} (\sqrt{2}\,V_{ph}) \cos \alpha = \frac{3\sqrt{3}}{\pi} \times \sqrt{2} \times \frac{V_1}{\sqrt{3}} \cos \alpha$$
$$= \frac{3\sqrt{2}}{\pi} V_1 \cos \alpha = 1.35\,V_1 \cos \alpha \qquad (4.15)$$

where V_1 is the line voltage.

4.14.1 Current waveforms of 6-pulse converter

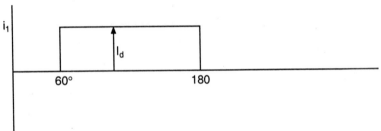

(h) Thyristor current

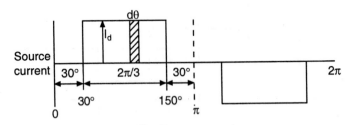

(i) Source current

Fig. 4.11

i_1 to i_6 are the currents through the SCRs. Each SCR conducts for 120°. Therefore thyristor current waveform has a block of 120° duration as shown in Fig. 4.11(g). It can be seen that source current is positive when odd numbered SCR conduct and source current is negative when even numbered SCR conduct. With this notation the source current waveforms i_R, i_Y and i_B can be drawn. It can be seen that the source current waveforms are 120° quasi square waves. Thus it can be concluded that the rectifier draws quasi square current from the source.

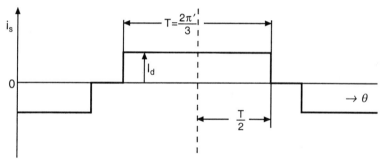

Fig. 4.11(j)

RMS value of SCR current in 6-pulse converter

The SCR current waveform is shown in Fig. 4.11(h). Limits are from 60° to 180°.

$$I_{RMS}^2 = \frac{1}{2\pi} \int_{60}^{180} I_d^2 \, d\theta = \frac{1}{2\pi} \int_{0}^{120} I_d^2 \, d\theta$$

$$= \frac{1}{2\pi} \int_{0}^{2\pi/3} I_d^2 \, d\theta = \frac{I_d^2}{2\pi} |\theta|_0^{2\pi/3} = \frac{I_d^2}{2\pi} \frac{2\pi}{3} = \frac{I_d^2}{3}$$

$$I_{RMS} = \frac{I_d}{\sqrt{3}} \tag{4.16}$$

RMS value of source current

The source current waveform is shown in Fig. 4.11(i). Limits are from 30° to 150°.

$$I_{sRMS} = \sqrt{1/\pi \int i^2 \, d\theta}$$

$$I_{sRMS}^2 = \frac{1}{\pi} \int I_d^2 \, d\theta = \frac{1}{\pi} \int_{30}^{150} I_d^2 \, d\theta = \frac{I_d^2}{\pi} \int_0^{120} d\theta$$

$$= \frac{I_d^2}{\pi} \int_0^{2\pi/3} d\theta = \frac{I_d^2}{\pi} |\theta|_0^{2\pi/3} = \frac{2\pi}{3} \times \frac{I_d^2}{\pi}$$

$$I_{sRMS}^2 = \frac{2I_d^2}{3}; \quad I_{sRMS} = \sqrt{2/3} \, I_d \tag{4.17}$$

RMS value of fundamental of I_s

The source current waveform shown in Fig. 4.11(j) has even mirror symmetry. Positive and negative areas of the source current are equal. Therefore there is no D.C. component in the source current. In the Fourier series for this waveform sine terms will be absent. Cosine terms will be present. a_n is given by the following equation.

$$a_n = \frac{1}{\pi/4} \int_0^{T/2} f(\theta) \cos n\theta \, d\theta; \quad a_1 = \frac{4}{\pi} \int_0^{60} f(\theta) \cos \theta \, d\theta$$

$$= \frac{4}{\pi} \int_0^{\pi/3} I_d \cos \theta \, d\theta = \frac{4I_d}{\pi} \int_0^{\pi/3} \cos \theta \, d\theta = \frac{4I_d}{\pi} |\sin \theta|_0^{\pi/3}$$

$$= \frac{4I_d}{\pi} [\sin \pi/3 - \sin 0] = \frac{4I_d}{\pi} \frac{\sqrt{3}}{2}$$

$$a_1 = \frac{2\sqrt{3} \, I_d}{\pi} \tag{4.18}$$

$$I_s = a_1 \sin \omega t + a_2 \sin 2\omega t + \ldots \tag{4.19}$$

$$I_s = i_{s1} + \ldots + \ldots \tag{4.20}$$

From (4.19) and (4.20) $i_{s1} = a_1 \sin \omega t$.
Substituting (4.18) here

$$i_{s1} = \left(\frac{2\sqrt{3} \, I_d}{\pi}\right) \sin \omega t; \quad I_{s1 \, RMS} = \frac{a_1}{\sqrt{2}} = \frac{2\sqrt{3} \, I_d}{\pi \sqrt{2}}$$

$$I_{s1 \, RMS} = \frac{\sqrt{6} \, I_d}{\pi} \tag{4.21}$$

Relation between power factor angle and firing angle

Rectifier converts A.C. power into D.C. power. 3ϕ A.C. input power is equal to D.C. output power if the losses in the devices are neglected.

In all the circuits fundamental component of the A.C. current produce useful power. Harmonics produce additional heat. Therefore
Fundamental AC power = DC power

$$3 \, V_{RMS} \, I_{S1RMS} \cos \phi = V_d I_d$$

$$3 \, V_{RMS} \left(\frac{\sqrt{6} \, I_d}{\pi}\right) \cos \phi = \left(\frac{3\sqrt{3} \, V_m}{\pi} \cos \alpha\right) I_d$$

$$V_{RMS} \sqrt{6} \cos \phi = \sqrt{3} \, V_m \cos \alpha$$

$$V_{RMS} \sqrt{6} \cos \phi = \sqrt{3} \cdot \sqrt{2} \, V_{RMS} \cos \alpha$$

$$\cos \phi = \cos \alpha; \quad \phi = \alpha \tag{4.22}$$

Power factor angle = Firing angle

4.15 SYNCHRONIZED UJT TRIGGERING CIRCUIT

The circuit is shown in Fig. 4.12(a). In phase controlled rectifiers, the firing angle delay is measured from the zero crossing of the supply voltage. The firing pulses to be given to the SCRs must be synchronized to the supply voltage since the delay is measured from the zero crossing of the supply. This can be ensured by deriving the supply for the control circuit from the supply

Phase Controlled Rectifiers 59

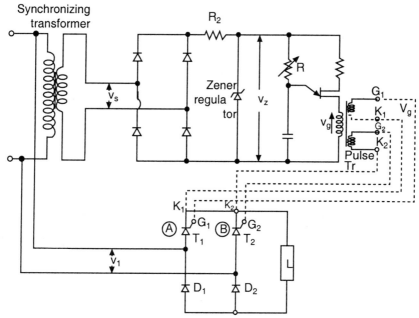

(a) Synchronized UJT Triggering Circuit

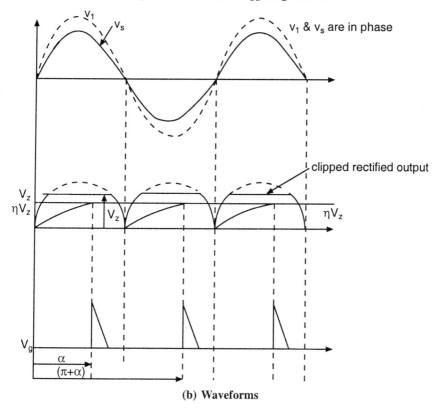

(b) Waveforms

Fig. 4.12

given to power circuit through a transformer. The low voltage a.c. is converted to full wave rectified output by using diode bridge rectifier. The resistance R_2 and zener diode maintains almost a constant voltage since the zener diode acts as a voltage regulator. The output across the zener diode is a clipped full wave rectified output. When the capacitor voltage is less than peak voltage, the UJT is OFF. The capacitor exponentially charges. When the capacitor voltage exceeds $\eta\ V_z$, UJT switch latches and capacitor discharges through the primary winding of the pulse transformer. The pulse transformer has two secondaries. Each of the secondary is connected to the gate and cathode of the SCRs. During the positive half cycle, the pulse is given to both the SCRs, but the SCR T_1 alone conducts since it is forward biased. Similarly during the negative half cycle, the SCR T_2 alone conducts. Thus the above circuit, generates train of synchronised pulses required for the rectifier circuit.

4.16 FIRING CIRCUIT FOR 3-ϕ CONVERTER

Block diagram of the firing circuit is shown in Fig. 4.13. The line voltage V_{RY} is stepped down using step down transformer. Output of transformer is given to the ZCD. Output of the ZCD is high for a duration of 180° and low for a duration of 180°. The wave form C is input to the saw-tooth generator. Saw tooth generator produces a saw tooth voltage for a duration corresponding to one half cycle. This is one of the inputs to the comparator. The other input is D.C. voltage or control voltage V_c. Output of the comparator is high for a duration where triangular voltage is greater than D.C. voltage. Output pulses of comparator 1 are given to the gate of SCR 1. Output of ZCD is inverted using an inverter. A saw tooth waveform is generated corresponding to the output of inverter (G). This is one of the inputs to the comparator 2. The other input is the control voltage. Output of comparator 2 is high for a duration where triangular voltage is greater than D.C. voltage. This output is given to the Gate of SCR 4.

Transformers are not required if the waveforms are shifted by 120° and 240° for getting pulses to the second and third legs. The blocks of control circuit for the pulses of second leg and third leg are similar to that of first leg. From the waveform it can be seen that the firing angle delay can be varied by varying the value of control voltage.

List of Formulae

Formulae for 3 Phase 6 Pulse Converter with Overlap
Let ΔV be the voltage drop in a 3ϕ rectifier due to the overlap.
    V_{du} → output voltage of 3ϕ converter considering overlap
    V_d → output voltage of 6 pulse converter without overlap
    $V_{du} = V_d - \Delta V$

(i) $V_{du} = \dfrac{3\sqrt{3}\ V_m}{2\pi} [\cos \alpha + \cos(\alpha + u)]$ (4.23)

where u = overlap angle; when $u = 0$

Phase Controlled Rectifiers

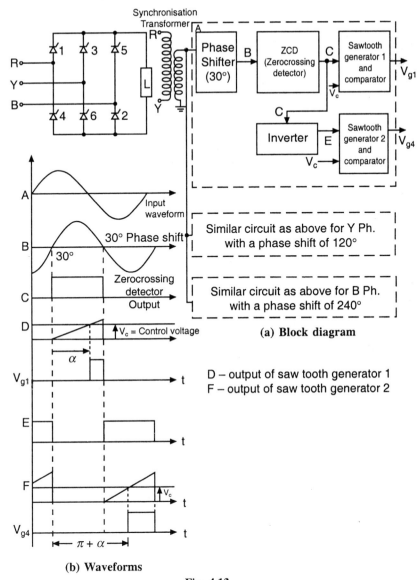

(a) Block diagram

D – output of saw tooth generator 1
F – output of saw tooth generator 2

(b) Waveforms

Fig. 4.13

$$V_d = \frac{3\sqrt{3}\, V_m}{\pi} (\cos \alpha)$$

Considering overlap

$$\cos \phi = \frac{\cos \alpha + \cos (\alpha + u)}{2} \qquad (4.24)$$

where u = overlap angle.
when $u = 0$, $\cos \phi = \cos \alpha$

Problems

Ex. 4.1 A 6 pulse converter connected to secondary of Δ/y, 6.6 KV/415 V, 50 Hz transformer is supplying 460 V, 200 A D.C. load. Calculate
(i) α
(ii) D.C. power
(iii) A.C. Line Current
(iv) R.M.S. value of device current

Solution
(i) $\qquad V_0 = 1.35\, V_1 \cos \alpha; \qquad 460 = 1.35 * 415 * \cos \alpha$

$$\cos \alpha = 0.82 \qquad \alpha = 34.9°$$

(ii) D.C. power $P_0 = V_0 I_d = 460 * 200 = 92$ kw

(iii) A.C. line current

$$I = \sqrt{2/3}\, I_d = \sqrt{2/3} * 200 = 163.3 \text{A}$$

(iv) RMS value of SCR current

$$I_{sRms} = \frac{I_d}{\sqrt{3}} = 200/\sqrt{3} = 173.2 \text{A}$$

Ex. 4.2 A 3ϕ full converter operates from 3ϕ, 415 V, 50 Hz. It feeds power to the armature of 440 V D.C. shunt motor. A regenerative braking is applied while motoring and the back e.m.f. being 420 V. Calculate alpha so as to restrict the regenerative current to 4.5A. Assume the effective loop resistance to be 0.7 Ω and neglect all other voltage drops

In regenerative braking, the polarity of e.m.f. is reversed by reversing the field connection. e.m.f. direction is reversed to reverse the direction of power. Current direction can not be reversed due to the presence of SCRs.

$$\text{By KVL, } V_0 = -E + I_a R_a; \qquad 1.35\, V_L \cos \alpha = -E + I_a R_a$$

$$1.35 * 415 * \cos \alpha = -420 + 4.5 * 0.7; \qquad \alpha = 133.9°$$

α is greater than 90° since the converter acts as an inverter.

Ex. 4.3 A 1ϕ fully controlled bridge converter with A.C. source voltage of 230 V feeds a load consisting of R, L and e.m.f. If $R = 0.1\ \Omega$ and rated current = 25A, calculate
(i) α, if $E = 120$ volts
(ii) α, if $E = -120$ volts
Assume continuous conduction.
(i) Using KVL

$$V_0 - I_a R_a - E = 0; \qquad V_0 = E + I_a R_a; \qquad \frac{2V_m}{\pi} \cos \alpha = E + I_a R_a$$

$$\frac{2\sqrt{2} * 230}{\pi} \cos \alpha = 120 + (25 * 0.1); \qquad \alpha = 53.7°$$

(ii)
$$V_0 = -E + I_a R_a$$

$$\frac{2V_m}{\pi} \cos \alpha = -E + I_a R_a; \qquad \frac{2\sqrt{2} * 230}{\pi} \cos \alpha = -120 + (25 * 0.1)$$

$$\alpha = 124.6°$$

Ex. 4.4 Derive an expression for ripple factor of half controlled rectifier feeding resistive load.

Phase Controlled Rectifiers 63

Ripple factor $(r) = \dfrac{\text{A.C. component in the output}}{\text{Average output}}$

$$r = \dfrac{\sqrt{V_{RMS}^2 - V_0^2}}{V_0} = \sqrt{\left(\dfrac{V_R}{V_0}\right)^2 - 1}$$

$$V_0 = \dfrac{1}{2\pi}\int_0^\pi V_m \sin\theta\, d\theta = \dfrac{V_m}{2\pi}[1 + \cos\alpha]$$

$$V_{RMS}^2 = \dfrac{1}{2\pi}\int_\alpha^\pi V^2\, d\theta = \dfrac{1}{2\pi}\int V_m^2 \sin^2\theta\, d\theta$$

$$= \dfrac{V_m^2}{2\pi}\int \dfrac{(1-\cos 2\theta)}{2}\, d\theta = \dfrac{V_m^2}{4\pi}\left[\int_\alpha^\pi d\theta - \int_\alpha^\pi \cos 2\theta\, d\theta\right]$$

$$= \dfrac{V_m^2}{4\pi}\left[(\pi-\alpha) - \left|\dfrac{\sin 2\theta}{2}\right|_\alpha^\pi\right] = \dfrac{V_m^2}{4\pi}\left[(\pi-\alpha) + \dfrac{\sin 2\alpha}{2}\right]$$

Therefore $\quad V_{RMS} = \dfrac{V_m}{2\sqrt{\pi}}\left[(\pi-\alpha) + \dfrac{\sin 2\alpha}{2}\right]^{1/2}; \; r = \sqrt{\dfrac{V_R^2}{V_0^2} - 1}$

Therefore $\quad r = \sqrt{\dfrac{\dfrac{V_m^2}{4\pi}(\pi-\alpha) + \dfrac{\sin 2\alpha}{2}}{\dfrac{V_m^2}{4\pi^2}(1+\cos\alpha)^2} - 1}$

$$r = \sqrt{\dfrac{\pi\left[\pi-\alpha + \left(\dfrac{\sin 2\alpha}{2}\right)\right]}{(1+\cos\alpha)^2} - 1}$$

Ex. 4.5 A single phase fully controlled bridge rectifier supplied from a 230 V single phase A.C. sources feeds a resistance load of 20Ω. Calculate the power dissipated in the load resistance when the rectifier operates at a firing angle of 45°.

$$P = \dfrac{V_{RMS}^2}{R}; \quad V_{RMS}^2 = \dfrac{1}{\pi}\int_\alpha^\pi V^2\, d\theta$$

$$V_{RMS} = \dfrac{V_m}{\sqrt{2\pi}}\left[(\pi-\alpha) + \dfrac{\sin 2\alpha}{2}\right]^{1/2} = \dfrac{\sqrt{2}*230}{\sqrt{2\pi}}\left\{\left[\pi-\dfrac{\pi}{4}\right] + \dfrac{\sin 90}{2}\right\}^{1/2}$$

$$= 219.3 \text{ volts}$$

$$P = \dfrac{V_R^2}{R} = \dfrac{(219.3)^2}{20} = 2404 \text{ watts}$$

Ex. 4.6 Calculate the largest average output voltage of a 3φ half controlled rectifier. The converter is fed from 400 V A.C. supply.

Solution

$$V_0 = \frac{3\sqrt{3}\,V_m}{2\pi} \cos\alpha = \frac{1.35\,V_1 \cos\alpha}{2}$$

If $\alpha = 0$, V_0 is maximum, therefore

$$V_0 = \frac{3\sqrt{3}\,V_m}{2\pi} = (1.35/2)\,V_1 = \frac{1.35}{2} * 400 = 270\ V$$

Ex. 4.7 A 3ϕ, 6 pulse fully controlled converter is connected to 3ϕ A.C. supply of 400 V and 50 Hz. It operates with the firing angle of 45°. The load current is maintained constant at 10 A and the output voltage is 360 V. Calculate the source inductance and overlap angle.

Solution

$$\omega = 2\pi f = 2\pi * 50 = 314$$

$$V_{du} = 1.35\,V_1 \cos\alpha - \frac{3I_0 X_L}{\pi}$$

$$360 = 1.35 * 400 * \cos 45° - \frac{3 * 10 * \omega L}{\pi}$$

We get $L = 7.38$ mH using (4.23)

$$360 = \frac{3\sqrt{3}\left(\dfrac{400\sqrt{2}}{\sqrt{3}}\right)}{2\pi}\,[\cos 45° + \cos(45 + u)]$$

$$u = 6.3°$$

$$R_L = \frac{V_0}{I_0} = \frac{360}{10} = 36\ \Omega;\quad \cos\phi = \frac{\cos\alpha + [\cos(\alpha + \mu)]}{2}$$

$$\cos\phi = \frac{\cos 45 + \cos(45 + 6.3)}{2} = 0.199$$

Ex. 4.8 A 3ϕ full fridge circuit is used for rectification. The leakage inductance of each phase of transformer winding is 2 mH. The 3ϕ input voltage is 400 V at 50 Hz. The load current on the D.C. side is 15 amps. Calculate
 (a) drop in the D.C. output voltage due to source reactance
 (b) firing angle required for the SCRs when D.C. output voltage is 200 V.

Solution

(a) Drop $= \Delta V = \dfrac{3I_0 X}{\pi}$

$$= \frac{3 * 15 * 314 * 2 * 10^{-3}}{\pi} = 9\ V$$

(b) $V_{du} = V_d - \Delta V = 1.35\,V_1 \cos\alpha - \Delta V$

$$200 = 1.35 * 400 * \cos\alpha - 9;\quad \alpha = 67.6°$$

Short Questions and Answers

1. Effect of source inductance in a controlled rectifier is to *reduce* output voltage.
2. In a single phase bridge rectifier, the PIV rating of the diode is *Peak value of supply voltage*.

3. A phase controlled converter may be operated as an inverter when the phase angle delay is *90 to 180°*.
4. In a semi-controlled bridge rectifier the power flow is from *source to load*.
5. Ripple factor of centre tapped full wave rectifier is *0.482*.
6. For the same average output voltage, PIV rating of SCRs will be larger for single phase *full wave centre tapped rectifier*.
7. Zig-Zag secondary of a rectifier transformer is to avoid *dc saturation*.
8. A single phase fully controlled converter is charging a battery from existing AC mains. It is possible to feed power to ac supply when α is between 90 and 180 with *reverse battery connection*.
9. A flywheel diode across the inductive load helps in *commutation process*.
10. So far as generation of triggering pules is concerned the use of cosine method is essentially *open loop control*.
11. A half controlled converter operates as controlled rectifier and inverter.
12. In an inductive/resistive circuit satisfactory triggering of SCR is done by an *astable* multi-vibrator.
13. The reading of ac ammeter connected to a half controlled rectifier with firing angle $\alpha = 0$ will be

$$I_{RMS} = \sqrt{\frac{1}{2\pi} \int_0^\pi i^2 \, d\theta} = \frac{I_m}{2}$$

14. Thyristor in the following circuit does not require commutation.
    Ans. Rectifier
15. Gate triggering of thyristor uses pulse transformer to provide *isolation*.
16. What is false triggering?
    Ans. Sequence in single phase circuit is 1 2, 3 4, 12, 34. sequence in 3 phase converter is 1-2, 2-3, 3-4 5-6, 6-1. Any sequence other than this is called false triggering.
17. Why pulse triggering is preferred. When does it fail?
    Ans. With multiple pules the SCR triggering is successful. Gate loss is reduced. Saturation of pulse transformer is not there. When there is a problem in the control circuit or firing circuit, it does not work.
18. What is the function of ZCD in the firing circuit?
    Ans. To convert sine wave into a square wave.
19. From what point the firing angle delay is measured in 3-ϕ converters?
    Ans. From the natural point of commutation which is 30° after 0° crossing.
20. A fully controlled converter feeds a DC machine. What are the conditions for regeneration of power?
    Ans. 1. $\alpha > 90°$
    2. Polarity of the armature voltage has to be reversed.
    3. AC system should be in a position to absorb the power.
21. What are the advantages of free wheeling diode?
    Ans. 1. power factor is improved.
    2. Average voltage is increased.
22. Why regeneration is not possible with semi converter?
    Ans. Polarity of the DC Output voltage cannot be reversed. Direction of current cannot be reversed due to the presence of unidirectional devices. Therefore regeneration is not possible.
23. In theory, why a free wheeling diode is not required for a semi converter?
    Ans. Freewheeling diode is not required since semiconverter has inherent freewheeling.

24. What are the effects of overlap?

Ans. 1. Power factor is reduced. 2. Average output voltage is reduced.

25. What is the frequency of ripple in the output of (a) single phase full wave rectifier (b) 3-ϕ half controlled rectifier (3 pluseconverter) (c) 6 pulse converter (or) 3-ϕ fully controlled rectifier?

Ans. (a) $2f$ (b) $3f$ (c) $6f$

(f = Supply frequency)

5

D.C. Choppers

5.1 INTRODUCTION

In many applications, DC voltage may have to be changed from one value to another. Transformers perform this task in AC systems very efficiently. In potential divider and D.C. motor generator sets, the efficiency is very low. It is possible to construct choppers which step up or step down DC voltages very efficiently.

5.2 BASIC PRINCIPLE

The basic principle of obtaining a variable dc voltage from a fixed dc voltage using solid state devices can be understood from the circuit given in Fig. 5.1(a) S is a mechanical switch which is opened and closed periodically. When the switch is closed, the load is connected across the battery and the voltage across the load is V_s. The switch is kept closed for a period t_{on}. Then the switch is opened and kept open for a period t_{off}. During this period, the load is disconnected from the battery and the voltage across the load is zero. The waveform of voltage across the load is shown in Fig. 5.1(b). From this waveform, the average value of dc voltage across the load can be obtained as

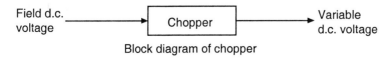

Block diagram of chopper

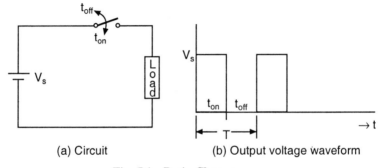

(a) Circuit   (b) Output voltage waveform

Fig. 5.1 Basic Chopper

68 Fundamentals of Power Electronics

$$V_{av} = \frac{V_s * t_{on}}{t_{on} + t_{off}} = \frac{V_s * t_{on}}{T} = V_s t_{on} f = V_s \delta \qquad (5.1)$$

where

$$\delta = \frac{t_{on}}{T} = t_{on} f$$

The quantity δ is called as duty cycle or time ratio of chopper, t_{on} is the on period of chopper and f is the frequency of chopper.

It may be noted that the output voltage is obtained by chopping the input voltage. Hence this circuit is called as a chopper. The time ratio of chopper can be varied from zero to one and correspondingly the average output voltage varies from 0 to V_s. If the output voltage is less than the input voltage, it is called as step down chopper.

5.3 CONTROL STRATEGIES

The time ratio control of the switch in Fig. 5.2 (a) can be done by the following methods:

(i) Variable frequency TRC control
(ii) Constant frequency TRC control

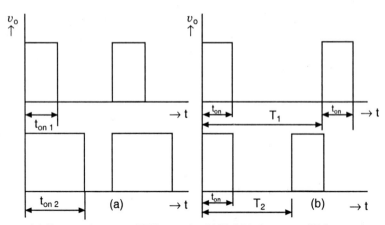

(a) Constant frequency TRC control   (b) Variable frequency TRC control

Fig. 5.2 Waveforms of output voltages for the two control strategies

Equation (5.1) has two variables t_{on} and f and among them one may be varied while the other is kept constant, to control the output voltage. In the case of variable frequency TRC control, t_{on} is kept constant while f is varied. In the case of constant frequency TRC control, f is kept constant while t_{on} is varied. The waveforms of the output voltage for these two methods of control are shown in Fig. 5.2(b). The chopper frequency is normally selected in the range of 300 Hz to 2 kHz. The higher the frequency, the smaller is the size of filter components and faster is the response. The constant frequency TRC control has an advantage that designing the filter becomes simple and economical since the frequency is constant.

The rms output voltage, $V_{rms} = \sqrt{\dfrac{V_s^2 \, t_{on}}{T}} = V_s \sqrt{\delta}$

The rms load current, $I_{rms} = \dfrac{V_{rms}}{R} = \dfrac{V_s \sqrt{\delta}}{R}$

Total output power, $P_t = V_{rms} \cdot I_{rms} = \dfrac{V_s \sqrt{\delta} * V_s \sqrt{\delta}}{R}$

$$P_t = \dfrac{V_s^2 \, \delta}{R}$$

The average output power, $P_{av} = V_s \, I_{av} = V_s \dfrac{V_{av}}{R}$

where $I_{av} = \dfrac{V_s \, \delta}{R}$ and $P_{av} = \dfrac{V_s^2 \, \delta}{R}$ (5.2)

5.4 CLASSIFICATION OF CHOPPERS

Practical choppers use solid state switching devices such as power transistors, thyristors or GTO instead of mechanical switch. Among the above devices, the most popular device used in high voltage and high current choppers is thyristor. Such choppers are mostly used for controlling the speed of dc motors. In a chopper, the thyristor may occupy different places in the actual power circuit of the motor. Based on the location and the on and off periods of the thyristor, the polarities of load voltage and load current change giving rise to different quadrants of operation in the V-I characteristic of the chopper. Apart from the quadrants of operation, the motor may operate in motoring mode drawing power from the battery or operate in generating mode delivering power to the battery. Based on the above operations, the choppers are classified in terms of quadrants and in terms of motoring/regenerating modes of load. A chopper needs a commutation circuit since the input voltage to the chopper is dc. It is also classified based on the class of commutation circuit used. Sometimes choppers are simply classified as Type A, Type B etc.

Type A Chopper

Fig. 5.3(a) shows a chopper with thyristor T_1 in series with the load. When T_1 is on, current flows from the battery to the load and the voltage across the load is V_s. when T_1 is off, the load current freewheels through the diode and the voltage across the load becomes zero. In this case during both the on and off periods of chopper, the current in the load remains in the same direction but the voltage changes between V_s and zero. So the operating points of the chopper lies in the first quadrant of V-I characteristic. This type of chopper is called Ist quadrant chopper or motoring chopper or simply Type A chopper.

Type B Chopper

In Fig. 5.3(b) the chopper has a thyristor T_2 in parallel with the load. When T_2 is on, the motor current circulates through T_2 and the voltage across motor

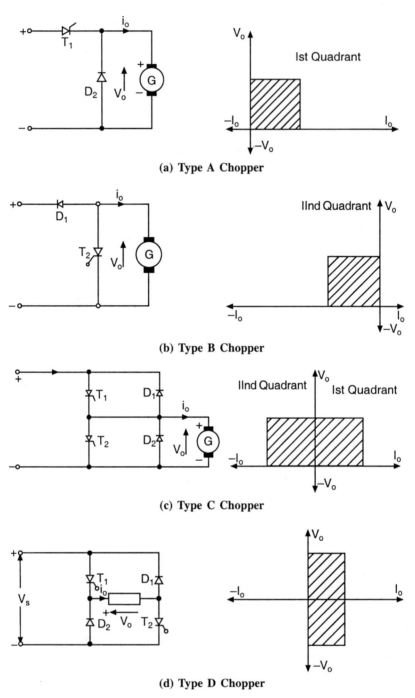

Fig. 5.3

becomes zero. It may be noted that the current through the motor is in the opposite direction to the current in Fig. 5.3(a). When T_2 is off, the voltage across motor is V_s and current flows into the battery through D_2. The voltage

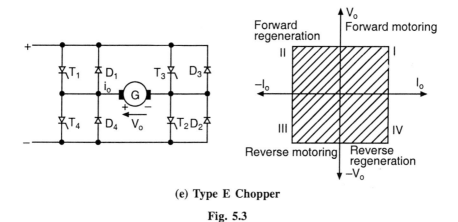

(e) Type E Chopper

Fig. 5.3

and current corresponds to the second quadrant of V-I characteristic. Hence, the chopper is called IInd quadrant chopper or regenerative chopper.

The chopper of Fig. 5.3(c) has thyristors T_1 and T_2. T_1 is in series and T_2 is in parallel with the load. The choppers are operated in complementary manner. When T_1 is on, load voltage is V_s and the load current is in the positive direction. The motor receives power from the battery. When T_1 is off, D_2 conducts and the load current freewheels through D_2. The freewheeling current is due to the energy stored in the inductance of the motor. After the current has become zero, T_2 is turned on. The load voltage still remains at zero whereas the load current reverses its direction. This current is due to the back emf of the motor. This type of chopper has operations in the first and second quadrants. This configuration of chopper is called as two quadrant chopper (1–2), or motoring and regenerative chopper.

Type D Chopper

The chopper shown in Fig. 5.3(d) has two thyristors and two diodes forming a bridge. When T_1 and T_2 are on, voltage across the load is V_s and the load current is in the positive direction. Power flows from the battery to the motor. When T_2 is off, load current freewheels through D_1 and T_1. On the other hand, when T_1 is off, it freewheels through T_2 and D_2. In either case of freewheeling, the load voltage is zero and the load current is positive. When both T_1 and T_2 are off, the load voltage is negative and the load current is fed back to the battery.

Type E Chopper

Figure 5.3(e) is called 4-quadrant chopper since it operates in 4 modes. In Ist quadrant T_1 and T_2 conduct. V_0 and I_0 are positive. Hence the machine operates in forward motoring. If the machine e.m.f is greater than the supply voltage with a polarity such that D_1 and D_2 are forward biased, the machine pumps current to the source. V_0 is +ve and I_0 is negative. Hence it lies in second quadrant i.e the machine operates in forward generation mode. If SCRs T_3 and T_4 are triggered, the actual current and actual voltage are opposite to the reference directions. V_0 is negative, I_0 is negative. The machine operates in reverse motoring mode. It lies in third quadrant. If the polarity of generated emf forward biases the diodes D_3 and D_4, the machine pumps current into the source. If generated emf is greater than supply voltage, V_0 is negative and I_0 is positive. The machine operates in the reverse generation mode.

Four types of step down choppers are discussed in the following sections. They are

    (i) Voltage commutated chopper
    (ii) Current commutated chopper
    (iii) Load commutated chopper
    (iv) Jones chopper

5.5 VOLTAGE COMMUTATED CHOPPER

Voltage commutated chopper (VCC) uses class D commutation circuit. The circuit and modes of VCC are shown Fig. 5.4. The chopper feeds a constant current load (highly inductive load). To start with the capacitor is precharged with top plate positive by closing the switch shown in the Figure. The operation can be explained using 4 modes.

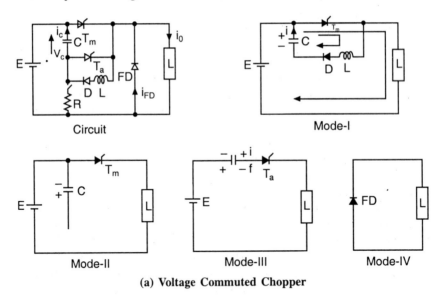

(a) Voltage Commuted Chopper

Mode 1 The main SCR T_m is triggered to start the ON period of the chopper. There are two current paths. The load current flows through main SCR and load. Ringing current flows through T_m – L-D and C. The ringing current is sinusoidal and capacitor voltage is co-sinusoidal. The total current through the main SCR is the sum of constant load current and ringing current.

Mode II The ringing stops after half a cycle. Later the load current alone flows through the SCR T_m. The polarity of the capacitor is with top plate negative. This is the required polarity for the commutation.

Mode III To initiate the OFF period of the chopper, the main SCR has to be turned off. When T_a is switched On, the capacitor pumps a large current through

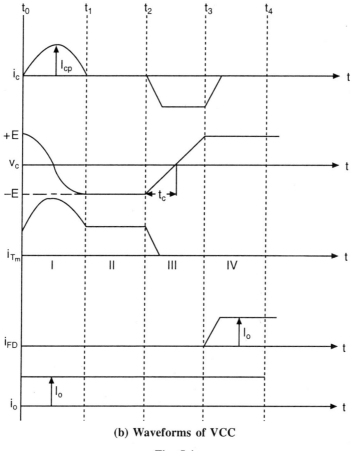

(b) Waveforms of VCC

Fig. 5.4

T_m and T_m turns OFF. Later the constant current flows through the capacitor, T_a and load. The capacitor charges linearly when a constant current flows through it.

Mode IV The voltage across the load inductance forward biases the freewheeling diode FD to maintain constant current through it. The current through T_A gradually decreases and current through FD gradually increases such that the sum of these currents are equal to the constant load current. When the current through T_a reduces to zero, it turns off naturally. FD alone conducts in this mode.

The values of L and C can be calculated using the following formulae

$$C = \frac{2t_q I_0}{E} \quad (5.3)$$

$$L = \frac{1.11 E^2 C}{I_0^2} \quad (5.4)$$

where t_q = Device sum off time, I_0 = Output current and E = Supply voltage.

5.5.1 Firing Circuit for VCC

Block diagram and waveforms of firing circuit are shown in Fig. 5.5(a) and 5.5(b) respectively. Saw tooth generator operates at a frequency equal to the output frequency of the chopper. The inputs to the comparator are saw tooth voltage and control voltage (V_C). Output of the comparator is high when $V_T > V_C$. Output is low when $V_T < V_C$. Output of the comparator is given to both rising and falling edge monostable multivibrators. The falling edge mono produces one pulse for every falling edge of the square wave (V_0). Rising edge mono

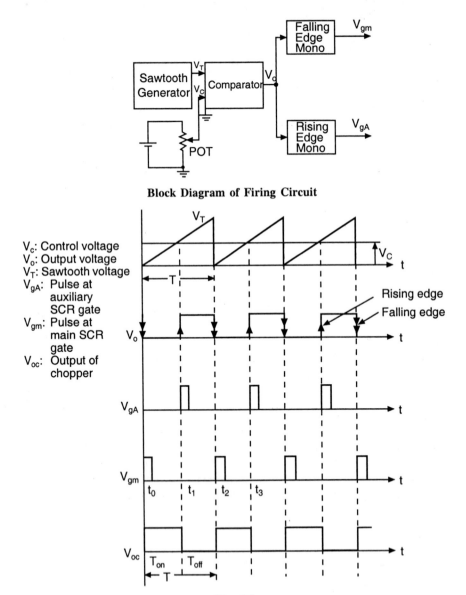

Fig. 5.5

gives a pulse corresponding to each rising edge of V_0. The output of falling edge mono is given to the gate of main SCR. Output of rising edge mono is given to the gate of auxiliary SCR (T_A). At $t = t_0$, a pulse is given to the main SCR (T_m), it conducts for a period T_{ON}. At $t = t_1$, a pulse is given to auxiliary SCR. This auxiliary SCR turns off the main SCR (T_m). At $t = t_2$ a pulse is again given to main SCR (T_m) and the above process repeats. If the control voltage increases, off period decreases and vice versa. Thus the average output voltage can be controlled by varying control voltage V_0.

5.6 CURRENT COMMUTATED CHOPPER (CCC)

The circuit and modes of CCC are shown in Fig. 5.6. In current commutated chopper circuit, an inductor is connected in series with the capacitor. T_m is the main SCR and T_A is the auxiliary SCR. To start with the capacitor is pre charged with top positive and bottom negative. The details of commutation can be explained with the following modes.

Mode I The main SCR (T_m) alone conducts and it causes the load current for the entire ON period of the chopper. The main SCR conducts for a period T_{ON}.

Mode II The commutation process starts with the triggering of auxiliary SCR (T_A). There are two current paths in the circuit. Load current flows through the source, T_m and load. Ringing current flows through T_A, L and C. Ringing takes place for half cycle. By the end of ringing, capacitor polarity gets reversed and current reduces to zero. Therefore the SCR T_A turns off.

Mode III After T_A is turned off, the circuit is formed by C, L, D_2 and T_m. The capacitor now drives a current through the SCR T_M in a direction opposite to the load current. The net current through the SCR is $i_0 - i_c$. From the wave

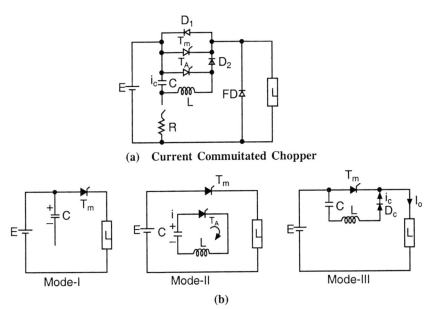

(a) Current Commuitated Chopper

(b)

76 *Fundamentals of Power Electronics*

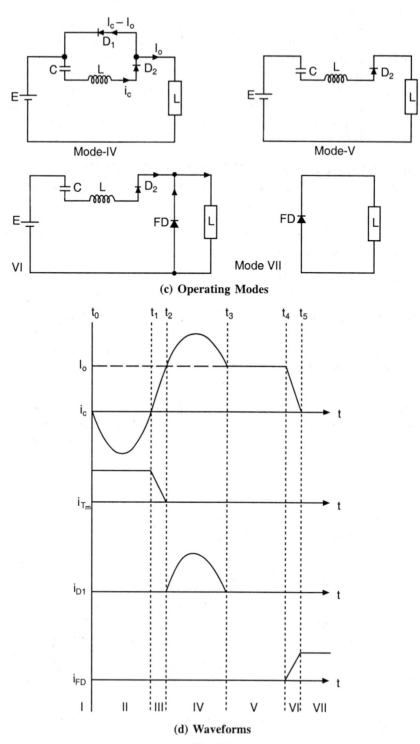

(c) **Operating Modes**

(d) **Waveforms**

Fig. 5.6

form, it can be seen that the instantaneous value of i_c is less than I_0 during this mode. At $t = t_2$, the value of $i_c = i_0$. Net current through the SCR is zero and the SCR T_m turns off.

Mode IV As long as T_m is conducting, the diode D_1 can not conduct since it is reverse biased by the drop across T_m. When T_m is turned off, the reverse bias on D_1 is removed and the ringing continues through D_1. The current through D_1 is $i_c - i_0$. At t_3 the instantaneous value of $i_c = i_0$. The diode D_1 turns off. During this mode, the SCR is reverse biased by the drop across D_1.

Mode V The constant current flows through the path E-C-L-D_2 and load. The current decreases beyond t_4. The load demands a constant current. The polarity of the voltage across the load inductor reverses to forward bias the freewheeling diode.

Mode VI The current through D_2 gradually decreases, while the current through free wheeling diode increases during mode VI.

Mode VII When the current through D_2 reduces to zero, D_2 turns off. The entire load current free wheels through the diode FD. If the main SCR is triggered again, the above modes get repeated. Successful commutation is possible only if the load current is less than the peak value of capacitor current.
The values of L and C can be calculated using the following formulae

$$L = \frac{V_s t_c}{x I_0 (\pi - 2\theta)} \quad (5.5)$$

where $x = \dfrac{I_{cp}}{I_0}$ and $\theta = \sin^{-1}(1/x)$.

$$C = \frac{x I_0 t_c}{V_s (\pi - 2\theta)} \quad (5.6)$$

where t_c = circuit turn off time and I_{cp} = peak value of capacitor current.

5.7 LOAD COMMUTATED CHOPPER

The load commutated chopper (LCC) uses four SCRs. They are triggered in pairs. Initially the capacitor is charged with a − and b +. The working of LCC can be explained with the following modes. The circuit and modes of LCC are shown in Fig. 5.7.

Mode I The SCRs T_1 and T_2 are triggered. The constant load current flows through T_1, C, T_2 and load. When constant current flows through the capacitor, the voltage varies linearly. The voltage across the load reduces from $2E$ to O. The current reduces to zero due to the under damped circuit formed by load resistance, load inductance and the capacitance C.

Mode II The polarity of voltage across the load inductance gets reversed to maintain the constant current through it. The entire load current flows through the diode FD.

78 Fundamentals of Power Electronics

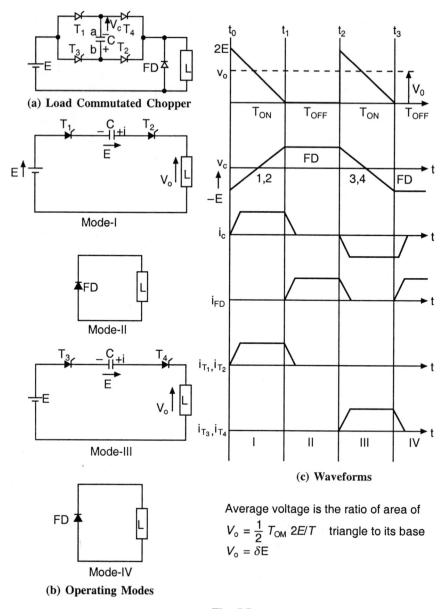

Fig. 5.7

Mode III When T_3 and T_4 are triggered simultaneously, the current transfers from the diode FD to the SCRs T_3 and T_4. The voltage across the load is zero during free wheeling.

The advantage of LCC is that it does not require a commutation inductor since load communication takes place. This is at the cost of two additional SCRs.

5.8 JONES CHOPPER

Jone's chopper does not require a resistance for precharging the capacitor. The capacitor is initially charged with plate a + by triggering auxiliary SCR alone. The working of Jones chopper can be explained with the following modes shown in Fig. 5.8. This circuit is similar to voltage commutated chopper except for an additional centre tapped reactor.

Mode I The main SCR T_M is triggered to have the on period of the chopper. There are two current paths in this mode. The load current flows through T_M, L_2 and load. Simultaneously a ringing circuit is formed by T_M, L, D_1 and C.

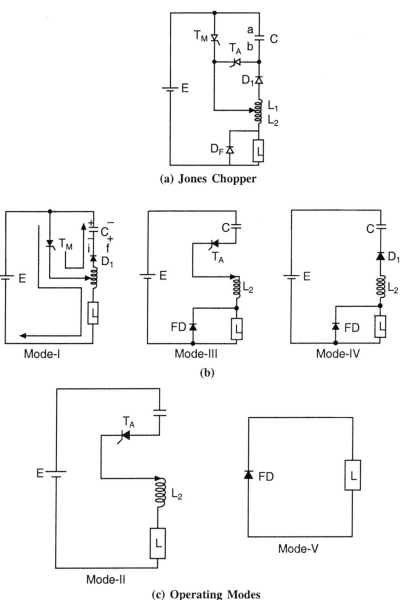

(a) Jones Chopper

(b)

(c) Operating Modes

Fig. 5.8

The ringing takes place for half a cycle. When the current reduces to zero, the diode D_1 turns off. The capacitor polarity reverses due to ringing. At the end of the on period of the chopper, the auxiliary SCR T_A is triggered to commutate T_M. The capacitor drives a large current through T_M and it gets turned off.

Mode II After T_M is turned off, the current continues to flow through C, T_A L_2 and load. The capacitor overcharges since it is charged through T_A and L_2.

Mode III The capacitor polarity forward biases the diode FD. The current through T_A decreases while the current through FD increases to maintain constant load current. When the current through T_A reduces to zero, it turns off.

Mode IV When T_A is conducting, D_1 cannot conduct since the drop across T_A reverses biases D_1. When T_A is off, the capacitor discharges through battery, FD, L_2, L_1, and D_1 until the capacitor voltage equals the supply voltage.

Mode V The load maintains constant current due to the conduction of the diode FD. Freewheeling alone takes place since all the devices are off.

It can be noted that the polarity of the capacitor in modes IV and V is similar to the initial polarity in mode I. If the main SCR is triggered again, the above modes get repeated.

The waveforms of this chopper are similar to those of VCC.

5.9 STEP-UP CHOPPER

Step up chopper is also called boost converter. The steady state wave form for current is shown in Fig. 5.9. At $t = t_0$, the SCR is turned on. A constant voltage is applied to the inductor.

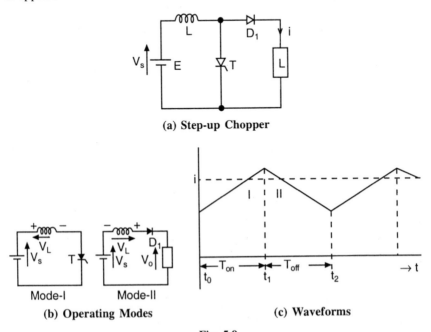

(a) Step-up Chopper

(b) Operating Modes

(c) Waveforms

Fig. 5.9

KVL for Mode 1 gives

$$V_L = V_s$$

$$V_s = V_L = L\, di/dt$$

V_s and L are constant. Therefore, di/dt has to be constant. Hence the current through the inductor increases linearly during the period T_{on}. After the SCR is turned off, the circuit is as shown in mode II. During this mode, the energy in the inductor is transferred to the load. The following assumptions are made in the derivation.

(i) The current is assumed to be constant or the ripple is neglected.

(ii) The energy given to the inductor during T_{on} is equal to the energy released to the load during T_{off}.

Let W_{on} and W_{off} be the energies corresponding to ON and OFF periods respectively.

$$W_{on} = V_s\, I\, t_{on} \tag{5.8}$$

Applying KVL in the circuit of mode II.

$$V_s + V_L - V_0 = 0$$

$$V_L = (V_0 - V_s)$$

Therefore
$$W_{off} = (V_0 - V_s)\, I\, t_{off} \tag{5.9}$$

Equating (5.8) and (5.9)

$$V_s\, I\, t_{on} = (V_0 - V_s)\, I\, t_{off}$$

$$V_s\, t_{on} = V_0\, t_{off} - V_s\, t_{off}$$

$$V_s\, (t_{on} + t_{off}) = V_0\, t_{off}$$

$$V_s\, T = V_0\, t_{off}$$

$$V_0 = V_s\, T/t_{off}$$

we know that

$$t_{on} + t_{off} = T$$

$$t_{off} = T - t_{on} = T - \delta T = (1 - \delta)\, T$$

Therefore
$$V_0 = V_s\, T/(1 - \delta)\, T$$

$$V_0 = V_s/1 - \delta \tag{5.10}$$

Since δ is always less then 1, output voltage is greater than input. Hence this circuit is called step up chopper.

Problems

Ex. 5.1. A separately excited d.c motor takes an armature current of 96 A on a 500 V supply and runs at 1000 r.p.m. $R_a = 0.25\ \Omega$. A chopper is used to control the speed of motor in the range 400 to 800 r.p.m. On time of the chopper is 2.5 msec. Determine the range of frequency of the chopper. The field is applied directly from 500 V supply.

Solution

$$E_0 = \delta E \text{ and Frequency} = \frac{1}{T}$$

Case I

$$N_1 = 400 \text{ r.p.m}$$

Back emf at 1000 r.p.m. $= V - I_1 R_a = 500 - 96*0.25$

$$E_{b1} = 476 \text{ volt}; \quad \frac{E_{b2}}{E_{b1}} = \frac{N_2 \phi_2}{N_1 \phi_1}$$

Since it is separately excited, ϕ is same.

Therefore $E_{b2} = (N_2/N_1) E_{b1} = [400/1000]\ 476 = 190.4$ volt

E_0 = Terminal voltage required for 400 r.p.m

$$= E_{b1} + I_a R_a = 190.4 + (96 * 0.25) = 214.4 \text{ volt}$$

To run at 400 r.p.m. voltage required is 214.4 volt

$$E_0 = \left(\frac{T_{ON}}{T}\right) E$$

$$214.4 = \frac{0.025}{T} * 500; \quad T = 5.84\ \mu\text{sec}$$

Frequency $\quad f = \dfrac{1}{T} = 171.2$ Hz

Case II

$$N = 800 \text{ r.p.m.}$$

$$E_{b1} = (N_2/N_1) E_{b1} = [\ 800/1000]\ 476 = 380.8 \text{ volts}$$

E_0 = Terminal voltage required for 800 r.p.m

$$= E + I_a R_a = 380.8 + (96 * 0.25) = 404.8 \text{ volt}$$

But $\quad E_0 = (T_{ON}/T)\ E$

$$404.8 = \frac{2.5 * 10^{-3}}{T} * 500; \quad T = 3.09 \text{ msec}$$

Therefore, frequency $= \dfrac{1}{T} = \dfrac{1}{3.09 * 10^{-3}} = 323.8$ Hz

Ex. 5.2. A 100 V separately excited d.c motor has armature and field resistance of 0.4 and 100 Ω. It takes an armature current of 25 A to drive a constant torque load at 1200 r.p.m. A chopper is used to control the speed of the motor. Find T_{ON} to reduce the speed to 800 r.p.m. at a chopper frequency of 500 Hz.

Solution Back emf at 1200 r.p.m. $= V - I_a R_a = 100 - 25 * 0.4$

$$E_{b1} = 90 \text{ volt}$$

$$E_{b2} = (N_2/N_1) E_{b1} = [800/1200]\ 90 = 60 \text{ volt}$$

Terminal voltage at 800 r.p.m. $= E_{b2} + I_a R_a = 60 + 25 * 0.4 = 70$ V

given $\quad T_1 = T_2$

or $\quad K\phi I_{a1} = k\phi I_{a2}; \quad I_{a1} = I_{a2}$

$$E_0 = \left(\frac{T_{ON}}{T}\right)E; \qquad T = \frac{1}{f} = \frac{1}{500} = 2 \text{ msec}$$

$$70 = \frac{T_{ON}}{2} * 100$$

$$T_{on} = 1.4 \text{ } \mu\text{sec}$$

Ex. 5.3. A voltage commutated chopper feeds power to a d.c motor. The d.c. input voltage is 60 V. The d.c. motor draws a constant current of 60 A. The device turn off time is 20μsec. Calculate the values of L and C.

Solution

$$C = \frac{2t_q I_0}{V_s} = \frac{2 * 20 * 10^{-6} 60}{60} = 40 \text{ } \mu f$$

$$L = 1.1 * \frac{V_s^2}{I_0^2} * C = 1.1 * \frac{60^2}{60^2} * 40 * 10^{-6} = 44 \text{ } \mu h$$

Ex. 5.4. For a current commutated chopper, peak value of capacitor current is twice the load current. The supply voltage is 230 V and t_q is 30 μsec. For a load current of 200 A Calculate (1) Peak value of capacitor current, (2) L (Commutation inductor) and (3) C (Commutation capacitor).

$$I_{cp} = 2I_0 = 2 \times 100 = 400 \text{ A}$$

$$L = \frac{V_s t_c}{x t_0 (\pi - 2\theta)}; \quad \frac{I_{cp}}{I_0} = 2 = x$$

$$L = \frac{230 * 2 * 30 * 10^{-6}}{2 * 200 (\pi - 2 * \pi/6)} = 13.1 \text{ } \mu\text{H}$$

$$C = \frac{X I_0 t_c}{V_s (\pi - 2\theta)} = \frac{2 * 200 * 60 * 10^{-6}}{230 (\pi - 2 * \pi/6)} = 39.8 \text{ } \mu\text{F}$$

Ex. 5.5. A Load commutated chopper fed D.C. drive uses 100 V D.C. supply. The maximum chopper frequency is 5 KHz. Calculate the value of commutating capacitance if the maximum load current is 100 A.

Solution

$$C = \frac{I_{om}}{2 E f_m} = \frac{100}{2 * 100 * 5000} = 100 \text{ } \mu\text{F}$$

Short Questions and Answers

1. What is the function of free wheeling diode in a chopper?
    Ans. (a) It protects SCR from high voltage that may be induced when the inductive circuit is interrupted.
    (b) It helps to maintain constant current through the load.
    (c) It helps to commutate main SCR.
2. Which power electronic circuit is dc equivalent of transformer?
    Ans. DC chopper converts fixed dc to variable dc. Auto transformer converts fixed ac into variable ac.

84 *Fundamentals of Power Electronics*

3. What is the disadvantage of frequency modulated chopper?
   Ans. Design of filter is difficult.
4. What is the function of diode in type *B* or regenerative chopper?
   Ans. It prevents short circuiting of dc source.
5. Why a commutation inductor is not required in LCC?
   Ans. Load inductance itself is used for commutation.
6. What is the function of diode in series with the inductor of VCC?
   Ans. This prevents current flow in opposite direction so that capacitor voltage remains with a polarity required for voltage commutation.
7. What are the applications of D.C. choppers?
   Ans. (a) Electric locomotives
   (b) Battery operated cars
   (c) Power supplies
8. What are the disadvantages of D.C. choppers?
   Ans. (a) They use forced commutation
   (b) They produce Electromagnetic Interference (EMI)
   (c) They have more switching losses and stresses

6
Inverters

6.1 INTRODUCTION
Inverters convert DC power into AC power of variable voltage and variable frequency. Important applications are induction heating, UPS system, HVDC system and speed control of AC motors. This chapter discusses some important types of inverter circuits.

6.2 SERIES INVERTER
The circuit, modes and waveforms are shown in Fig. 6.1. This is called series

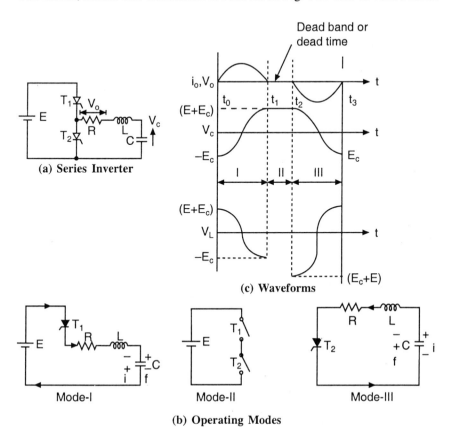

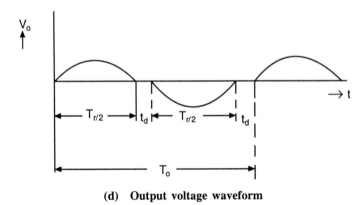

(d) Output voltage waveform

Fig. 6.1

inverter since the load resistance is in series with L and C. R is the load resistance. L and C are the commutating elements. This inverter uses two SCRs. The operation is as follows.

Mode I SCR T_1 is triggered at t_0. The capacitor is precharged with top negative. R, L and C form an under damped circuit and a sinusoidal current flows through the load resistance. When the current reduces to zero, the SCR T_1 turns off. The voltage across load resistance is in phase with the current through it. The wave forms V_L and V_C can be obtained by applying KVL ($v_L + v_C = E$). v_L and v_C have to be drawn to satisfy this equation.

Mode II The SCR T_2 should not be triggered immediately after the current through T_1 reduces to zero. A reverse bias should be applied to T_1 for successful commutation. If T_2 is triggered without time delay or dead band, the battery gets shorted through T_1 and T_2. None of the devices conduct in this mode. $V_R = 0$, $V_L = 0$ since $L\, di/dt = 0$. V_c remains unchanged.

Mode III At t_3, the SCR T_2 is triggered to initiate the negative half cycle. The capacitor discharges through L, R and T_2. It can be seen that the current through R is in opposite direction. When this current reduces to zero, the SCR T_2 turns off. The waveforms can be obtained by applying KVL ($v_L + v_c = 0$). Waveforms of v_L and v_c are drawn such that $v_L + v_c = 0$.
After a dead band, if T_1 is triggered, the above modes get repeated.

6.2.1 Advantages
1. Simple circuit
2. Output is nearly sinusoidal

6.2.2 Disadvantages
1. L and C are bulky.
2. Battery is utilised only during the positive half cycle.
3. The output voltage has harmonics due to the dead band.

Series inverter is best suitable for high frequency applications since the values of L and C required are very small.

From Fig. 6.1d,
Time period for one cycle = $T_o = T_r + 2t_d$

$$f_0 = \frac{1}{T_r + 2t_d} \quad (6.1)$$

where $T_r = 1/f_r$ and t_d is dead time.

$$f_r = \frac{1}{2\pi}\sqrt{\frac{1}{LC} - \frac{R^2}{4L^2}} \quad (6.2)$$

The output frequency is always less than resonant frequency due to dead band. Output frequency can be varied by varying the dead band.

6.3 PARALLEL INVERTER

Basic circuit of parallel inverter is shown in Fig. 6.2a. When switch 1 is closed, dot end of sections A, B & C is positive. Output voltage is positive. After T/2 seconds, open switch 1 and close switch 2. Dot ends of sections A, B and C are negative, output voltage is negative.

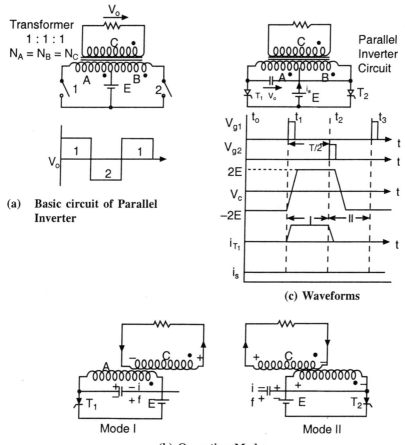

(a) Basic circuit of Parallel Inverter

(c) Waveforms

(b) Operating Modes

Fig. 6.2

The circuit, modes and waveforms are shown in Fig. 6.2. For low frequency applications, parallel inverters are used. It uses a centre tapped transformer, two SCRs and commutation capacitor. The dc source is connected between the centre tap and the common cathode point. The load resistance referred to the primary comes in parallel with the commutation capacitor. Hence this inverter is called parallel inverter.

The capacitor is precharged to $2E$ with a positive and b negative. At $t = t_1$, the SCR T_1 is triggered. The voltage E is applied across the section A of the 1 : 1 : 1 transformer. By the principle of self induction, an equal amount of voltage is induced in the section B with dot end positive and non dot end negative. The total voltage across the primary is $2E$. Therefore capacitor voltage reverses and reaches $+2E$. The polarity on the secondary side is positive at the dot end. The current in secondary flows through the resistance in anticlockwise direction.

At $t = t_2$, the SCR T_2 is triggered to turn off T_1. The final polarity of mode I is such that it pumps a large current through SCR 1. It gets turned off quickly by the principle of voltage commutation. The current in the primary flows through the path E, section B and T_2. The voltage polarity at the secondary is negative at the dot end. The load current flows in the clockwise direction in the secondary. Thus an alternating square current flows in the load resistance. Output voltage waveform is similar to capacitor voltage.

Disadvantages
1. Capacitor has to be rated for 2E.
2. The source current is not a pure dc current.
3. The ripple in the source current produce additional heating in the battery.

6.4 BRIDGE INVERTERS

1-φ half bridge Inverter
1-φ Half bridge inverter uses two sources and two switches. The load is connected between the centre tap of the source and centre point of the switches.

6.4.1 R-Load
The circuit and waveforms are shown in Fig. 6.3. The SCR T_1 conducts for a period $T_0/2$ ($T_0 = 1/f_0$). SCR 1 is commutated at $T_0/2$ and then SCR T_2 is triggered to initiate the negative half cycle at the load. SCR T_2 is commutated

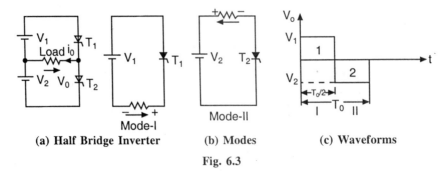

(a) Half Bridge Inverter   (b) Modes   (c) Waveforms

Fig. 6.3

at $t = T_0$. This process is repeated to obtain a continuous square wave. It must be ensured that T_1 and T_2 are not triggered simultaneously.

6.4.2 L-Load

The operation can be explained with the following four modes. The diodes D_1 and D_2 are called feed back diodes. The inverter cannot handle inductive loads without feed back diodes. Without D_1 and D_2, if SCR is commutated, a large voltage is induced across the load inductance. This voltage may damage the SCRs. The operating modes and waveforms are shown in Fig. 6.4.

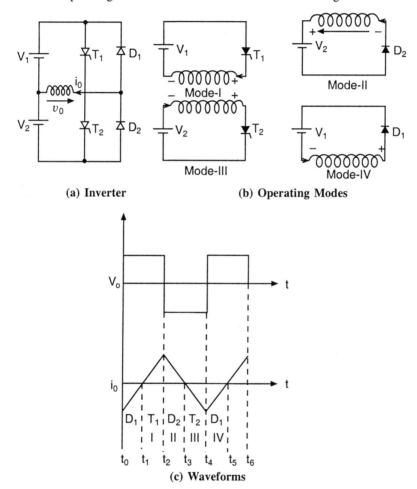

Fig. 6.4 1φ Half Bridge Inverter With L-Load

Mode I T_1 conducts to have the positive half cycle at the load. The current through the inductance rises linearly. At $t = t_2$, SCR T_1 is force commutated. The load inductance reverses its polarity to maintain current in the same direction.

Mode II The polarity across the load forward biases the diode D_2. D_2 conducts and power is pumped to the source V_2. when the current reduces to zero, the diode D_2 turns off.

Mode III As long as D_2 is conducting, T_2 cannot conduct since it applies a reverse bias across T_2. Once D_2 is off, T_2 conducts when it is triggered. In this mode both v and i are negative and power is positive i.e., the power flows from source to the load. At $t = t_4$, SCR T_2 is force commutated.

Mode IV The load inductance reverses its polarity to maintain the current in the same direction. This polarity forward biases diode D_1. Recycling takes place since the current flows into V_1. This happens till the entire energy is pumped to the source $V_1 \cdot D_1$ turns off at t_5. If T_1 is triggered, the above modes get repeated.

With RL load, the current varies exponentially. The positive area and negative area are not equal since some power is consumed by the load resistance in every cycle.

6.4.3 Half Bridge Inverter with RLC Load

The circuit and waveforms are shown in Fig. 6.5. When the inverter feeds RLC load, a separate commutation circuit is not required. This can be explained by using the waveforms shown in Fig. 6.5. The frequency of operation is such that $X_c > X_L$. Therefore the power factor is leading. The overall circuit is under damped. The current varies sinusoidally. From t_0 to t_1, the SCR T_1 conducts. At $t = t_1$, T_1 turns off since the current reduces to zero. From t_1 to t_2, the diode D_1 conducts and pumps the power to the source. Diode D_1 conducts since it is forward biased by the capacitor voltage. When D_1 conducts, it applies reverse bias to the SCR T_1. Thus a separate forced commutation circuit is not required. This circuit uses load commutation since the load itself commutates the SCRs.

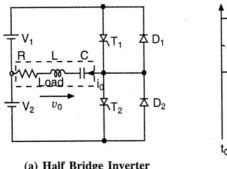
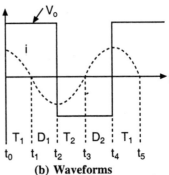

(a) Half Bridge Inverter (b) Waveforms

Fig. 6.5

In the negative half cycle T_2 conducts, D_2 follows T_2 and applies the reverse bias to the SCR T_2.

6.5 Mc-MURRAY INVERTER (VOLTAGE SOURCE INVERTER)

The Mc-Murray Inverter uses current commutation. A half bridge inverter feeding an inductive load is shown in the Fig. 6.6. T_{A1} and T_{A2} are the auxiliary SCRs. They are used for commutating the main SCRs T_1 and T_2. The elements L and C are the commutating components. The capacitor is precharged with left negative and right positive. The operation is explained using the following modes.

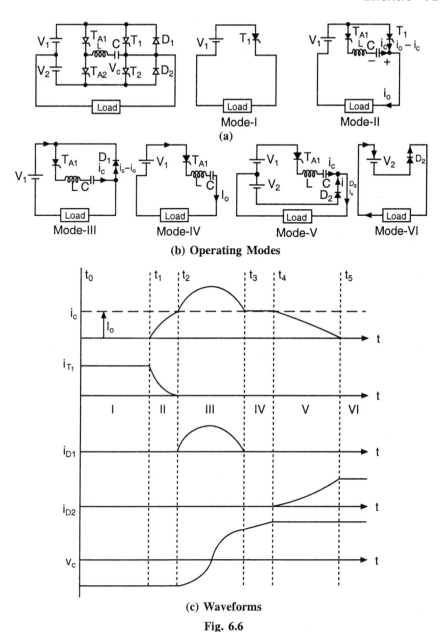

Fig. 6.6

Mode I SCR T_1 is triggered to initiate the positive half cycle. Constant load current flows through T_1.

Mode II Auxiliary SCR T_{A_1} is triggered. The elements L, C, T_1 and T_{A_1} form a ringing circuit. The capacitor current increases sinusoidally as can be seen from the waveform. In the duration t_1 to t_2, the value of $i_c < I_0$. At $t = t_2$, $i_c = I_0$. The current through SCR T_1 is zero and it turns off. During this mode it can be seen that the current through SCR T_1 reduces to zero.

Mode III After the SCR T_1 is turned off, the ringing continues through the diode D_1. Diode conducts till t_3 since $i_c - I_0$ is positive during this mode. At $t = t_3$, the diode D_1 turns off since its current reduces to zero.

Mode IV After the diode D_1 is turned off, the constant load current flows through the capacitor and it is further charged with left positive and right negative. The capacitor voltage varies linearly since a constant current flows through it.

Mode V The diode current increases while the capacitor current decreases. The same can be observed in the waveforms. When the current through T_{A_1} reduces to zero, it turns off.

Mode VI The energy stored in the load inductance forward biases the diode D_2. Recycling takes place and the energy in the load is pumped into the source V_2. After D_2 is turned OFF, the SCR T_2 is triggered.

To commutate T_2, TA_2 is triggered. Similar modes get repeated and the commutation process is same as above.

L and C can be calculated using the following formulae

$$C = \frac{0.89\, I_{om}\, t_c}{V_{min}} \tag{6.3}$$

$$L = \frac{0.39\, t_c\, V_{min}}{I_{om}} \tag{6.4}$$

where I_{om} = maximum output current and V_{min} = minimum output voltage.

Inverter should be designed for worst conditions like minimum input voltage and maximum output current.

6.6 Mc-MURRAY-BEDFORD INVERTER

Mc-Murray inverter uses two auxiliary SCRs. Mc-Murray-Bedford Inverter doesnot require any auxiliary SCRs. One main SCR can commutate the other main SCR. The operation is explained by using the following modes. Operating modes and waveforms are shown in Fig. 6.7.

Mode I SCR T_1 is triggered. Constant current flows through T_1 and L_1. Voltage across L_1 is zero since constant current flows through it. C_1 is shorted by T_1 and L_1. The capacitor C_2 charges to $V_1 + V_2$ with top + and bottom –ve.

Mode II The SCR T_2 is triggered to commutate T_1. When T_2 is triggered, the voltage across C_2 appears at L_2. This voltage is 2V. By the principle of self induction, an equal amount of voltage is induced in L_1. The cathode of T_1 is at 4 V and anode is at 2 V. The SCR T_1 is turned off and reverse bias is applied by the centre taped inductor. Quick turn off (T_1) is possible since the energy in the section L_1 is transferred to L_2 by constant flux linkage theorem. From the waveforms it can be seen that the current is transferred instantaneously from T_1 to T_2 in the beginning of mode II. L_2 and C_2 forms a ringing circuit.

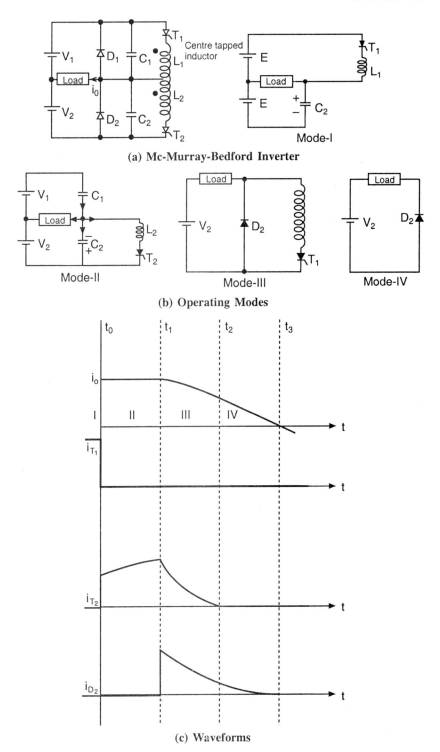

(a) Mc-Murray-Bedford Inverter

(b) Operating Modes

(c) Waveforms

Fig. 6.7

The diode D_2 is reverse biased by the voltage across C_2.

Mode III Once the capacitor polarity reverses, the diode D_2 gets forward biased. When D_2 is conducting, C_2 gets shorted by D_2. The energy trapped in L_2 maintains a current through T_2 and D_2. The energy is dissipated in the resistance of the coil L_2, resistance of T_2 and D_2. When the entire energy is dissipated, T_2 gets turned off.

Mode IV D_2 is forward biased by the load inductance. Recycling takes place and D_2 conducts until the energy in the load is pumped to the source V_2.

T_2 is again triggered to have the negative half cycle. If T_1 is triggered to commutate T_2, modes similar to the above modes get repeated.

6.7 3-ϕ INVERTERS

There are two modes of operation in 3-ϕ inverters. They are (i) 120° mode (ii) 180° mode.

6.7.1 120° Mode

A 3-phase bridge inverter has three legs. The SCRs are numbered in a way similar to 3-phase, 6 pulse converter. The difference of numbers in each leg is equal to three. A 3-ϕ Bridge Inverter feeding star connected resistive load is shown in Fig. 6.8. In 120° mode of operation, each SCR conducts for a duration of 120° in a cycle. At any time, two SCRs conduct in the circuit. Therefore two loads are energised. When the SCR from the odd group conducts, the respective node has positive potential. When the SCR from the even group conducts, the respective node has negative potential. The phase voltages are 120° quasi square waves. The line to line voltages are six-stepped voltage waveforms. Phase voltages are displaced by 120°. Line to line voltages are also displaced by 120° as shown in Fig. 6.8. The SCRs are triggered in the sequence 61–12–23–34–45–56. Output frequency can be varied by varying triggering frequency.

6.7.2 180° Mode

In 180° mode of operation, each SCR conducts for a duration of 180°. The two possibilities are as follows. Two devices from the odd group and one from the even group or two from the even group and one from odd group conduct. The potential of the node is positive if the odd group devices conducts. It is negative if even group devices conducts. At any time two resistances come in parallel and the third resistance comes in series with them. The magnitude of voltage is $V/3$ across the parallel branches and $2V/3$ across the series branch. The phase voltages are six-stepped waveforms and line to line voltages are 120° quasi square waves. They are shown in Fig. 6.9. Phase voltages are displaced by 120°. Line voltages are also displaced by 120°. The SCRs are triggered in the sequence 561–612–123–234–345–456.

6.8 3-ϕ CURRENT SOURCE INVERTER

The circuit and the operating modes are shown in Fig. 6.10. This inverter is called K. Phillips inverter. This uses voltage commutation. When a large inductance is connected in series with a voltage source it acts as a current

Inverters 95

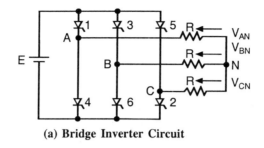

(a) Bridge Inverter Circuit

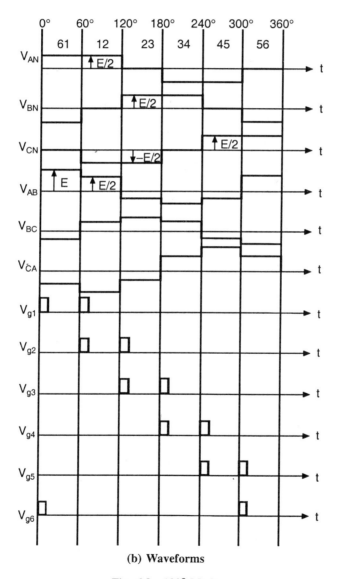

(b) Waveforms

Fig. 6.8 120° Mode

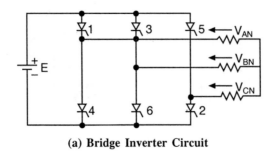

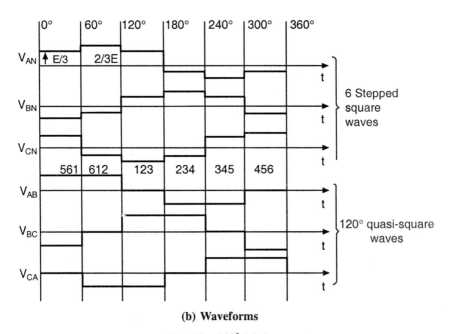

(b) Waveforms

Fig. 6.9 180° Mode

source. This operates in 120° mode operation. Six capacitors are required to turn off six SCRs. The diodes prevent discharging of capacitors into the load. The diodes D_1–D_6 are called isolation diodes. The SCRs are triggered in the sequence 12–23–34–45–56–61. Let us consider the change over from 12 to 23: SCR2 continues to conduct and conduction transfers from T_1 to T_3.

Mode I The capacitor C_1 is charged with left +ve and right −ve. The SCRs 1 and 2 are triggered as per 120° mode. The circuit remains in this state for 60°.

Mode II In the next 60° interval, 3 and 2 have to conduct. T_3 is triggered at the beginning of the 60° interval. T_1 gets turned off by voltage commutation. The current flows through D_1, A phase and C phase. The polarity of capacitor C_1 gets reversed.

Mode III D_1 continues to conduct since the load impedance in A phase

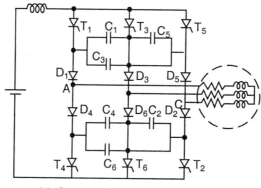

(a) Current Source Inverter

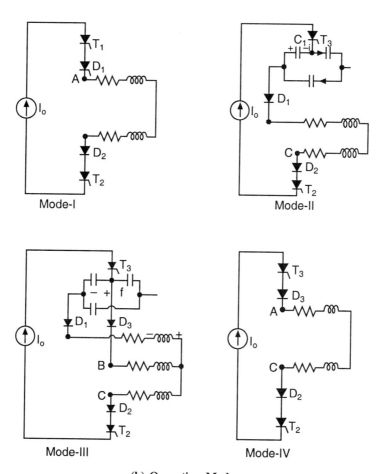

(b) Operating Modes

Fig. 6.10

98 *Fundamentals of Power Electronics*

demands current in the same direction. D_3 conducts since it is forward biased by the final polarity of the capacitor C_1. In this mode all three loads are energised. This is called the period of overlap.

Mode IV The diode D_1 conducts until the energy in the inductance of A-phase reduces to zero. Later the devices T_2 and T_3 conduct as per the regular sequence of 120° operation.

The phase current waveforms are similar to the phase voltage waveforms of VSI operating in 120° mode.

6.9 VOLTAGE CONTROL

Output voltage variation in the inverter is required in applications like speed control, UPS etc. The voltage control may be achieved using three methods.

1. Voltage control at the input
2. Voltage control at the output of the inverter
3. Voltage control within the inverter

The voltage at the input side can be controlled by having a phase controlled converter or chopper at the input. The disadvantage of phase controlled converter is that it operates at low power factor at the input side. The disadvantage of dc chopper is that they have high switching losses.

The ac voltage at the output can be controlled by having a transformer with tap changing gear. The disadvantages of tap changing are maintenance and sparking.

The voltage control within the inverter is called pulse width modulation. The types of PWM are: (1) Single PWM and (2) Multiple PWM.

6.9.1 Single PWM

The circuit and waveforms of SPWM are shown in Fig. 6.11. The output of single pulse width modulated inverter has one pulse per half cycle. The output

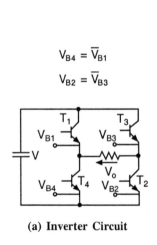
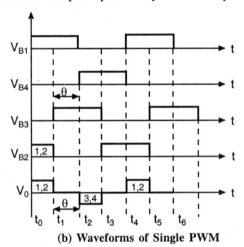

(a) Inverter Circuit (b) Waveforms of Single PWM

Fig. 6.11

voltage is varied by varying the width of the pulse in each half cycle. The base drive pulses of single PWM are shown in the figure 6.11b. The output voltage is available only if the transistors T_1 and T_2 (or) T_3 and T_4 conduct simultaneously. Only in the duration t_0 to t_1, the transistors T_1 and T_2 have base drive. They conduct in this duration and a positive pulse appears across the load. From t_2 to t_3, T_3 and T_4 are driven. Therefore a negative pulse appears across the load. The output voltage can be controlled by varying the angle θ. Larger the value of θ, smaller the output voltage and vice versa. The disadvantage of this method is that large harmonics are present in the output due to the single pulse.

6.9.2 Multiple Pulse Width Modulation

In multiple PWM, there are multiple pulses in each half cycle. Two types of multiple PWM are (i) Equal PWM (ii) Sine PWM.

(a) Equal PWM

The waveforms of symmetric PWM or equal PWM are shown in Fig. 6.12(a). Let V_T be the triangular voltage, V_C the control voltage and V_0 be the output voltage.

In the control circuit, a high frequency carrier voltage (triangular voltage) is compared with the control voltage (square wave form). The inputs to the comparator are V_T and V_C. The output of the comparator is high when V_T is more than V_C. Otherwise the output is low. The output of the comparator has a train of pulses. These pulses can be used to drive the power transistors. If it is a thyristorised inverter (the Murray Inverter), the main SCR is triggered at the beginning of the pulse and auxiliary SCR is triggered at the end of the pulse. Thus the output of multiple PWM inverter will have multiple pulses per half cycle. The harmonics of this waveform will be much less than that of the full square wave. The output of the comparator is used to drive the transistors of inverter shown in Fig. 6.11a.

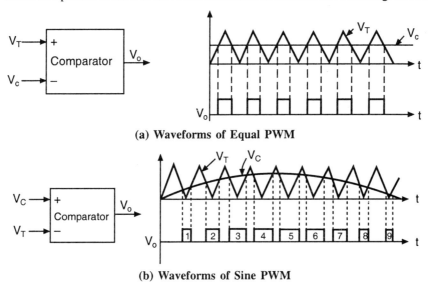

(a) Waveforms of Equal PWM

(b) Waveforms of Sine PWM

Fig. 6.12

(b) Sine PWM

The waveforms of sine PWM are shown in Fig. 6.12(b). In this modulation, the triangular waveform is compared with sinusoidal waveform. The input to the comparator are V_C and V_T. The output of the comparator is high when the magnitude of sinusoidal voltage is greater than the magnitude of triangular voltage. The ratio of control voltage to the triangular voltage is defined as modulation index. It can be seen that the output voltage has a train of pulses of unequal width. The width is maximum for the central pulse and it decreases on either sides. The width of the pulse varies sinusoidally. This is called asymmetric PWM since the width of the pulses are unequal. The harmonics of this voltage waveform will be less than equal PWM output.

6.10 HARMONIC CONTROL (WAVEFORM CONTROL)

The output voltage of the inverters may be a square wave, quasi square wave or six stepped wave or triangular wave. The output voltage will have fundamental and the associated harmonics. When the output of the inverter is fed to the induction motor, the harmonics produce additional heating. 5th harmonic voltage produces a torque in the opposite direction to the fundamental torque. Therefore it is preferred to reduce the harmonics in the output. The following methods are used to reduce the harmonics.

1. Tap changing
2. Transformer connections
3. Filters
4. PWM

6.10.1 Tap Changing

The solid state tap changing circuit is shown in Fig. 6.13(a). the circuit configuration is similar to a parallel inverter. When the SCR on the left side conducts the output voltage is positive. When the SCR on the right side conducts the output voltage is negative. When SCR 1 is triggered, the dc voltage is applied across half the primary of the transformer. The volt/turn is less with the conduction of SCR-1. The output voltage is also less. The SCR-1 is commutated and SCR-2 is triggered. The volt/turn is moderate and output voltage is moderate. When SCR-3 conducts, the volt/turn is high and the output voltage is high. The SCRs are triggered in the order 1-2-3-2-1-1A-2A-3A-2A-1A to get 12 stepped output. The disadvantage of the circuit is the requirement of complicated firing and commutation circuits.

6.10.2 Transformer Connections

Harmonic elimination using transformers is shown in Fig. 6.13(b). The output voltage in this system is obtained by using two inverters and two transformers. This method is used for eliminating a particular harmonic voltage in the output (Selective harmonic elimination). The secondary windings of the two transformers are connected in series aiding so that $V_1 + V_2 = V_0$. The firing pulses to the SCRs of the second inverter are delayed by θ. The waveform V_0 can be obtained by summing the waveforms V_1 and V_2. The output voltage is a 120° quasi square wave. For $\theta = 60°$, the vector diagrams for the fundamental and the third harmonics are shown in Fig.6.13(c).

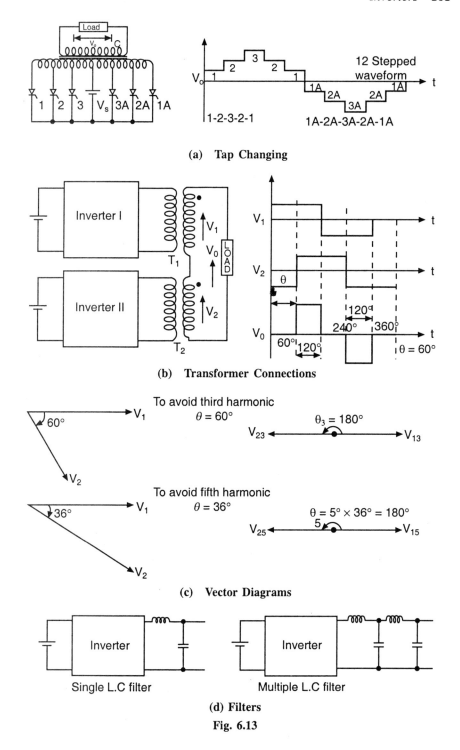

(a) Tap Changing

(b) Transformer Connections

(c) Vector Diagrams

(d) Filters

Fig. 6.13

The IIIrd harmonic voltages are anti phase with each other. Therefore the third harmonic does not appear in the output. By selecting $\theta = 36°$, the Vth harmonic in the output can be eliminated. The disadvantage of this system is the requirement of two inverter systems and two similar transformers.

6.10.3 Filters

Various types of filters used are shown in Fig. 6.13(c). In single LC filter, the inductors L_1 offers high reactance for higher frequencies. The high frequency components are dropped across L_1. The capacitor C_1 offers high reactance for low frequencies. By selecting a large L_1 the drop increases and the output voltage reduces on load. By choosing large C_1, high frequency components can be shunted through C_1. This increases the current rating of the inverter devices.

With single LC filter some harmonics manage to pass to the load. The harmonics can be further reduced by going for multiple filters. The size of filter inductor can be reduced by connecting it across the secondary of a step down transformer.

If the inverter operates at fixed frequency, series LC filter can be used. The values of L and C are chosen such that their natural frequency is equal to the output frequency of the inverter. The filter and load resistance operate as series resonant circuit. The current is in phase with the output voltage. Therefore the voltage across the load resistance is sinusoidal. This system is preferred for high frequency applications.

6.10.4 PWM

The output of PWM inverter has half wave symmetry. Positive area and negative area are equal. Therefore even harmonics are not present in the output. Odd harmonics can be reduced by selecting 10 pulses per half cycle. The higher order harmonics are permitted since they do not create any problem. The magnitude of higher order harmonics is very less. The lower order harmonics can be eliminated by using multiple PWM.

Problems

Ex. 6.1 A Mc-Murray inverter is fed from 50 V DC supply. The source may fluctuate by 4%. The turn off time of the SCR used is 25 μsec. The capacitor used is 50 μf. Determine the maximum current that can be commutated.
Solution

$$V_{min} = 50 - (4\% \text{ of } 50) = 50 - 2 = 48 \text{ V}$$

$$C = \frac{0.89 \, I_{om} t_c}{V_{min}} = \frac{0.89 * I_{om} * 50 * 10^{-6}}{48} = 50 \times 10^{-6}$$

Therefore $I_{om} = 54$ A.

Ex. 6.2 A Mc-Murray inverter is fed from a D.C source of 230 V. The D.C source may fluctuate by ± 20%. The current during commutation may vary from 10 A to 100 A. The turn off time of the device is 25 μsec. Calculate the values of 'L' and 'C'. Use a factor of safety of 2.

Solution
Factor of safety 2 means, $t_c = 2tq$.

$$C = \frac{0.89 \times 100 \times 50 \times 10^{-6}}{184} = 24.1 \, \mu f$$

Therefore $V = 230 - (20\% \text{ of } 230) = 230 - 46 = 184V$, $I = 100A$

$$L = \frac{0.39 \, t_c \, V_{\min}}{I_{om}} = \frac{0.39 \times 50 \times 10^{-6} \times 184}{100} = 35.8 \, \mu h$$

Short Questions and Answers

1. For variable voltage variable frequency application, inverter should supply power at *Variable voltage and variable frequency.*
2. How do you select SCR for inverters?
    Ans. (a) Turn OFF time should be minimum. Use inverter grade SCRs for low frequency application. Use BJT or MOSEFT for high frequencies.
    (b) Use devices with high dv/dt ratings.
    (c) Use factor of safety of 2 to 3.
3. In a certain circuit, 3 SCRs are triggered sequentially, later one SCR is triggered at a time. What is the mode of this inverter? *Ans.* 180° Mode.
4. In a voltage source inverter feedback diodes are used to pump the power to the source.
5. Commutation technique used in Mc-Murray Bedford inverter is voltage commutation.
6. A constant current source supplies a current of 100 mA to a load of 1K. When the load a changed to 100 ohms, what is the current? *Ans.* 100 mA
7. As far as harmonics in the output are concerned, multiple PWM provides less harmonics.
8. As compared to thyristor PMW inverter, the harmonics in the output voltage of transistor PWM inverter have *less* harmonics in BJT inverter.
9. PWM technique in inverter is used for harmonic reduction and voltage control.
10. What is function of feedback diodes in bridge inverter?
    Ans. Feedback diodes pump reactive current to the dc source.
11. Series inverter is suitable for *high* frequencies.
12. In sine PWM, the modulation index is kept < 1 to *reduce* the harmonics in the output.
13. In a basic series inverter the maximum output possible frequency can be below resonant frequency due *to dead band.*
14. What is the simplest method of eliminating 3rd harmonic in a 3-ϕ square wave inverter? *Ans.* Use 120° mode or connect the system in star.
15. The triggering signals for a class *C* commutated inverter are usually derived from Q and Q bar of a flip flop to *avoid* simultaneous triggering.
16. Discuss the advantages and disadvantages of transistor inverter over thyristor inverter?
    Ans. Advantages
    (1) High frequency operation is possible. (2) Fast response and less harmonics. (3) Commutation circuit is not required.
    Disadvantages
    (1) It requires a driving circuit. (2) Continuous base pulse should be applied since BJT does not have self latching property.

7

Cycloconverters

7.1 INTROUDCTION

Fixed voltage and fixed frequency supply can be converted into variable voltage, variable frequency supply using rectifier inverter system. The disadvantages of this system are requirement of two converters and increased losses.

The conversion of fixed voltage, fixed frequency supply to the variable voltage and variable frequency can be done without d.c. link by using cycloconverter. Cycloconverters are classified as (i) step down cycloconverter (ii) step up cyclyconverter. In a step down cycloconverter the output frequency is less than the input frequency. In a step up cycloconverter, the output frequency is greater than input frequency.

7.2 SINGLE PHASE STEP DOWN CYCLOCONVERTER

A step down cycloconverter has back to back connected full converters as shown in Fig. 7.1. At any time only one converter is in operation. When a P-converter operates, positive rectified segments appears across the load. When the N-converter operates, negative rectified segments appear across the load. Form the waveforms it can be observed that the output frequency is 1/3rd of the input frequency. Smaller firing angle delay for the middle pulse and larger delay for the other two pulses is chosen to reduce the harmonics in the output.

By selecting two pulses per half cycle, output frequency i.e., 1/2 the input

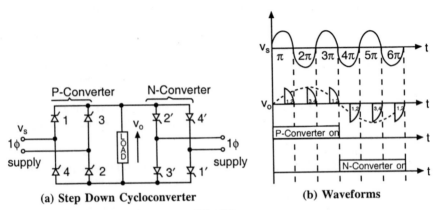

(a) Step Down Cycloconverter   (b) Waveforms

Fig. 7.1

frequency can be obtained. Thus the output frequency can be varied by varying the number of pulses per half cycle.

The output voltage can be varied by varying the firing angle delay of the converters.

DISADVANTAGES

1. Output voltage contains harmonics.
2. Smooth variation of output frequency is not possible.
3. A complicated firing circuit is needed.
4. Output frequency is an integer fraction of input frequency.

7.3 STEP-UP CYCLOCONVERTER

The circuit and waveforms of step up cycloconverter are shown in Fig. 7.2. The output frequency of step up cycloconverter is higher than the input frequency. This circuit uses two sets of antiparallel SCRs and a centre tapped transformer.

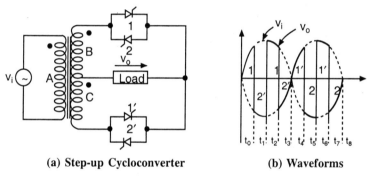

(a) Step-up Cycloconverter   (b) Waveforms

Fig. 7.2

During the positive half cycle, the dots assume positive polarity. The SCRs 1 and 2' are forward biased. At $t = t_0$, the SCR 1 is triggered. At t_1, SCR1 is force commutated and 2' is triggered. This is commutated at t_2 and again SCR1 is triggered. At t_3, SCR 1 is turned off and 2' is triggered.

During the negative half cycle, all the dots assume negative polarity. The SCRs 1' and 2 are forward biased. Similar process is repeated and SCRs are made to conduct in the order 1'-2-1'-2.

It can be seen that the output voltage waveform completes 4 cycles when the input waveform completes one cycle. Thus the frequency of output is four times the input frequency. The disadvantage of this circuit is the requirement of forced commutation.

7.4 3-ϕ TO 1-ϕ CYCLOCONVERTER

The circuit shown in Fig. 7.3 converts the 3-ϕ supply in to 1-ϕ supply. The above configuration is nothing but the dual converter configuration. At any time, firing pulses are given to only one of the converters. During the positive half cycle of the output, firing pulses are given to the P-convertor. During the negative half cycle of the output, firing pulses are given to the N-converter. The output of 6 pulse converter has six pulses per cycle of the input voltage.

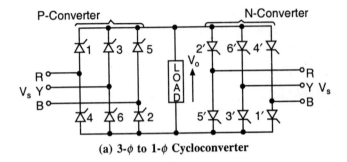

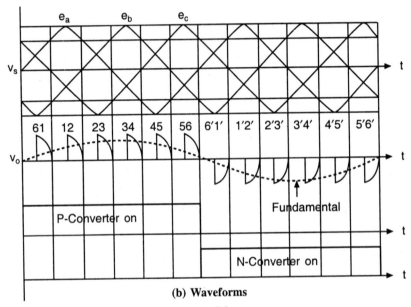

(b) Waveforms

Fig. 7.3

When N converter conducts, a negative voltage appears across the load. From the waveform, it can be seen that fundamental of the output completes one cycle when the input voltage completes two cycles. Therefore the frequency of the output voltage is half of the frequency of input voltage.

Disadvantages are the requirement of large number of thyristors and presence of harmonics in the output.

7.5 3-φ TO 3-φ CYCLOCONVERTER

The circuit shown in Fig. 7.4 uses three back-to-back connected converters. Each block is a three phase three pulse converter. The back-to-back converter system converts 3-φ voltage into 1-φ voltage by using the system explained above (3-φ to 1-φ).

The firing pulses for the system II are delayed by 120° with respect to the system I so that the voltage of phase B lags by 120° with respect to phase A. The pulses to the system III are delayed by 120° with respect to system II.

The firing circuit for the system is complicated since it uses 18 SCRs. This

Cycloconverters 107

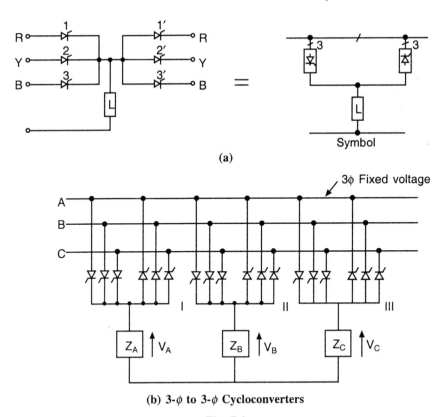

(a)

(b) 3-ϕ to 3-ϕ Cycloconverters

Fig. 7.4

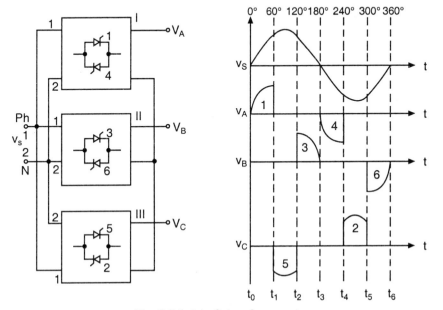

Fig. 7.5 1-ϕ to 3-ϕ cycloconverter

circuit is used to control the speed of 3-ϕ induction motor. The induction motor operates at low speed since low frequency supply is given.

7.6 1-ϕ TO 3-ϕ CYCLOCONVERTERS

In electric traction, the power is transmitted by using 1-ϕ system. In the locomotive, 1-ϕ supply can be converted to 3-ϕ supply and this 3-ϕ supply can be given to 3-ϕ induction motor. This system can be used instead of a phase controlled rectifier and DC motor system.

The circuit shown in Fig. 7.5 uses 3 a.c. choppers. The positive SCR in the A-phase is triggered at t_0. It is force commutated at t_1. The positive SCR of B phase is triggered at t_3. It is force commutated at t_4. Similarly the SCRs in B-phase and C-phase are triggered with a delay of 120° and 240° with respect to the A-phase SCRs.

The input terminals to the ac chopper of phase C are reversed to get proper 3-ϕ voltage at the output. It can be seen that 3 voltages displayed by 120° can be produced by properly chopping the input voltage. The output voltages have to be filtered before applying them to the load.

Short Questions and Answers

1. What is the relation between input and output frequencies of a step down cycloconverter?
   Ans. In a step down cycloconverter, the output frequency is an integer fraction of input frequency. $f_0 = f_i/n$
2. What is the disadvantage of step up cycloconverter?
   Ans. It requires forced commutation and output voltage contains harmonics.
3. Why 3-phase to 3-phase cycloconverter is not popular?
   Ans. It requires eighteen SCRs. The firing circuit is complicated. Hence 3-phase to 3-phase cycloconverter is not popular.
4. What are the advantages of cycloconverter?
   Ans. (i) They convert fixed voltage, fixed frequency supply into variable voltage, variable frequency supply without QC link.
   (ii) Conduction losses are lesser.

8

A.C. Choppers

8.1 INTRODUCTION

AC choppers convert fixed voltage, fixed frequency AC into variable voltage fixed frequency AC. These choppers are also called AC voltage regulators or AC bidirectional choppers. Some of the applications are heating, welding, starting and speed control of induction motors. This chapter discusses AC choppers with R and R-L load.

Figure 8.1 shows half controlled 1-ϕ ac voltage regulator. The circuit does not have control over the current in the negative half cycle. The output voltage does not have half wave symmetry. Hence even harmonics are also present.

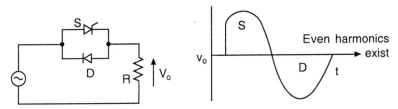

Fig. 8.1 AC Voltage Regulator

Figure 8.2 shows single phase bidirectional chopper using one SCR. During the positive half cycle, D_1 and D_2 conduct along with S provided it is triggered. During the negative half cycle, D_3 and D_4 conduct along with S. The voltage drop and conduction losses are more in this circuit since two diodes conduct along with SCR each time.

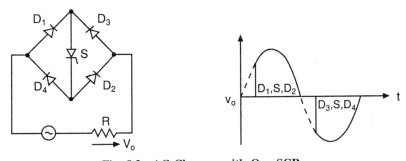

Fig. 8.2 AC Chopper with One SCR

8.2 AC CHOPPER USING TRIAC

The AC chopper circuit is shown in Fig. 8.3. Since the circuit chops the waveform on positive and the negative sides, it is called AC or bidirectional chopper. In this circuit, the pulse to the gate of triac is given through a diac. During the positive half cycle of the capacitor voltage, the capacitor has plate a positive and b negative. The diac remains off till the capacitor voltage is less than break over voltage V_{B1}. When capacitor voltage increases beyond V_{B1}, the diac breaks down and it acts as a closed switch. Now the capacitor voltage is applied between G and MT_1. The triac conducts and the voltage across the load is a segment of the supply voltage from α to π. Triac turns off at $\theta = \pi$ since the current reduces to zero.

In Fig. 2.2b, it is proved that v_c lags v_s by θ_1 in a series R-C circuit.

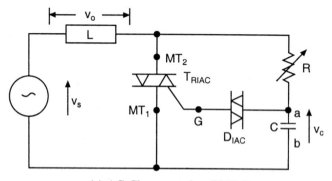

(a) AC Chopper using TRIAC

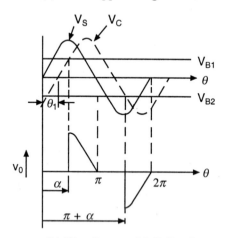

(b) Waveforms with R-Load

Fig. 8.3

During the negative half cycle, the capacitor has plate b positive and a negative. The diac turns on when the capacitor voltage is more than the break down voltage V_{B2}. When the diac conducts, the negative capacitor voltage is applied between G and MT_1. The triac conducts even for the negative voltage between gate and MT_1.

The voltage across the load is segment of the supply voltage. The firing angle delay can be varied by varying the resistance R or capacitance C. In this circuit, it is varied by using a potentiometer. It can be seen that the volt-second area of the output waveform is large when α is small and vice versa. Thus the output voltage can be varied by varying the firing angle.

Advantages
1. Less power loss.
2. Smooth control of output voltage.

Disadvantages
1. Harmonics are present in the output waveform since output voltage is not a pure sine wave.
2. This circuit introduces electromagnetic interference (EMI).

8.3 AC CHOPPER WITH R-LOAD

The waveforms for AC chopper with R load are shown in Fig. 8.4 At $\theta = \alpha$ the SCR T_1 is triggered. T_1 conducts up to π. The output current and

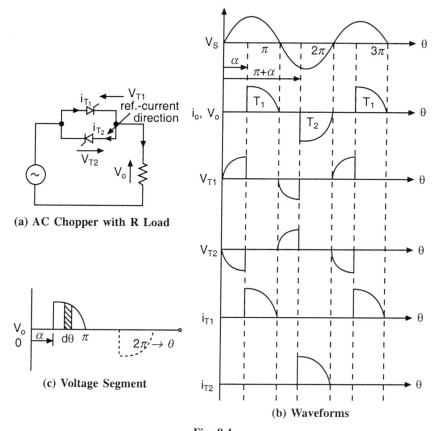

Fig. 8.4

output voltage wave forms are in phase since the load is resistive load. At $\theta = \pi$, T_1 turns off naturally since the current goes to zero. At $\theta = \pi + \alpha$, T_2 is triggered and it conducts up to 2π. This process is repeated for obtaining AC chopped waveform.

When either of the SCRs conduct, the voltage across them is zero ideally. When both of them does not conduct, the voltage across the switches is replica of the supply voltage.

8.3.1 R.M.S Value of the Output Voltage

The heat produced by the load in positive half cycle is same as the heat produced in negative cycle since the areas are equal. Hence, it is sufficient if we integrate from α to π by taking a base of π. From Fig. 8.4c.

$$V_{rms}^2 = \frac{1}{\pi} \int_\alpha^\pi v^2 \, d\theta = \frac{1}{\pi} \int_\alpha^\pi V_m^2 \sin^2 \theta \, d\theta$$

$$= \frac{V_m^2}{\pi} \int \sin^2 \theta \, d\theta = \frac{V_m^2}{\pi} \int \frac{(1 - \cos 2\theta) \, d\theta}{2}$$

$$V_{rms}^2 = \frac{V_m^2}{2\pi} \int (1 - \cos 2\theta) \, d\theta = \frac{V_m^2}{2\pi} \left[\left. \theta \right|_\alpha^\pi - \left. \frac{\sin 2\theta}{2} \right|_\alpha^\pi \right]$$

$$= \frac{V_m^2}{2\pi} [(\pi - \alpha) - (0 - \sin 2\alpha)/2]$$

Take root on both sides

Therefore $\qquad V_{rms} = \dfrac{V_m}{\sqrt{2\pi}} [\pi - \alpha) + (\sin 2\alpha)/2]^{1/2}$ \hfill (8.1)

Similarly $\qquad I_{rms} = \dfrac{I_m}{\sqrt{2\pi}} [(\pi - \alpha) + (\sin 2\alpha)/2]^{1/2}$ \hfill (8.2)

where $\qquad I_m = \dfrac{V_m}{R}$

8.4 AC CHOPPER WITH R-L LOAD

The waveforms with R-L load are shown in Fig. 8.5. Let ϕ be the impedance angle of the load. The min value of firing angle with RL load is ϕ. If SCRs are triggered with $\alpha > \phi$, controlled output voltage can be obtained. If SCRs are triggered with $\alpha = \phi$, full sine wave is applied across the load.

At $\theta = \alpha$, SCR-1 is triggered. It conducts up to β. From π to β, the forward bias to the SCR 1 is maintained by the inductive load. The period from α to π is called power period. The period from π to β is called recycling period. By increasing the firing angle, the width of the notch can be increased and vice versa.

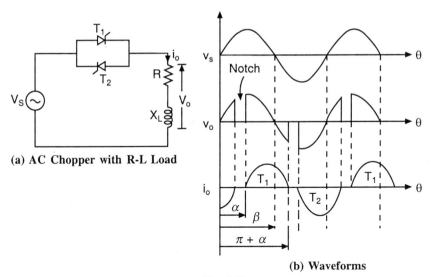

Fig. 8.5

Problems

Ex. 8.1 Two SCRs connected back to back have a load resistance of 400 Ω and a supply voltage of 110 V A.C. If the firing angle is 60°, find the R.M.S. output voltage.

Solution

$V_{rms} = 110$ volt; $R = 400\ \Omega$; $\alpha = 60°$

$$V_{rms} = \frac{V_m}{\sqrt{2\pi}}[(\pi - \alpha) + (\sin 2\alpha)/2]^{1/2}$$

$$V_{rms} = \frac{\sqrt{2}*100}{\sqrt{2\pi}}[(\pi - \pi/3) + (\sin 120)/2] = 88.5 \text{ volt}$$

Ex. 8.2 Prove that the form factor of the SCR current for an AC chopper feeding resistance load is

$$\frac{\sqrt{\pi}}{1+\cos\alpha}[(\pi - \alpha) + (\sin 2\alpha)/2]^{1/2}$$

Solution

Average current $= \frac{\text{Area}}{\text{Base}} = \frac{1}{2\pi}\int i\,d\theta = \frac{I_m}{2\pi}\int_\alpha^\pi \sin\theta\,d\theta = \frac{I_m}{2\pi}(1+\cos\alpha)$

$i_{Rms}^2 = \frac{1}{2\pi}\int_\alpha^\pi i^2 \cdot d\theta = \frac{I_m^2}{2\pi}\int_\alpha^\pi \sin^2\theta\,d\theta$

$i_{Rms} = \left[\frac{I_m^2}{2\pi*2}[(\pi-\alpha)+(\sin 2\alpha)/2]\right]^{1/2} = \frac{I_m}{2\sqrt{\pi}}[(\pi-\alpha)+(\sin 2\alpha)/2]^{1/2}$

Form Factor $= \frac{\text{R.M.S. value}}{\text{Average value}} = i_{RMS}/I_{av} = \frac{\sqrt{\pi}}{1+\cos\alpha}[(\pi-\alpha)+(\sin 2\alpha)/2]^{1/2}$

Ex. 8.3 A single phase A.C. regulator with 2 SCRs connected back to back has a resistive load of 10 Ω and a source voltge of 230 volt, 50 Hz; α- can be varied form 0 to 180°. Calculate (a) Greatest R.M.S. value and average value of SCR current. (b) Minimum circuit form off time (c) Maximum value of di/dt occurring in the SCR.

Solution

$$R = 10 \ \Omega \qquad V = 230 \ V$$

(a) Largest value of R.M.S. and average currents can be obtained with $\alpha = 0$ (because the area is more if $\alpha = 0$)

$$i_{T \ rms} = \frac{I_m}{2\sqrt{\pi}} [(\pi - \alpha) + (\sin 2\alpha)/2]^{1/2}$$

But

$$I_m = \frac{V_m}{R} = \frac{\sqrt{2}V}{R} = \frac{\sqrt{2} * 230}{10} = 23 * \sqrt{2}$$

with $\alpha = 0$,

$$i_{T \ RMS} = \frac{I_m}{2\sqrt{\pi}} (\pi)^{1/2}$$

$$i_{TRMS} = \frac{\sqrt{2} * 23}{2\sqrt{\pi}} (\pi)^{1/2} = 16.27 \ A$$

Substitute $\alpha = 0$ in I_{TAV}

$$I_{TAV} = \frac{I_m}{2\pi} (1 + \cos \alpha) = \frac{23\sqrt{2}}{2\pi} (1 + 1) = 10.3 \ A$$

(b) Minimum turn off time corresponds to maximum conduction period of SCR. The SCR can conduct for a maximum period of 180°. The time period corresponding to the remaining 180° is 10 msec.

$$\frac{1}{50} = 20 \ \text{msec}; \quad \frac{1}{2} * \frac{1}{50} = 10 \ \text{msec}$$

half the time period is 10 msec for frequency of 50 Hz.

(c) When $\theta = \alpha$, the value of $di/dt = \infty$ because $\tan 90° = \infty$
At $\theta = \alpha$, di/dt is maximum. Theoretically the value is ∞.

Short Questions and Answers

1. *Triac* may replace SCRs in *AC chopper* circuits.
2. What are the applications of AC chopper?
   Ans. (a) Speed control of ac motor.
   (b) Heating (electric).
   (c) Welding.
   (d) Reactive power control.
3. Do you require isolation between gating signals for two thyristors in a single phase full wave ac voltage controller?
   Ans. Yes.

9

Applications

9.1 INTRODUCTION
The two major areas of applications of semiconductor devices are motor and non-motor control applications. This chapter briefly discusses both these applications.

9.2 SPEED CONTROL OF INDUCTION MOTOR
The speed of induction motor can be varied by the following methods:
(a) Voltage Control
(b) V/f Control
(c) Rotor Resistance Control
(d) Slip Power Recovery Scheme

9.2.1 Voltage Control
We know that $N = N_s(1 - s)$

$$N = \frac{120f}{P}(1 - S); \quad T = \frac{KSE_2^2 r_2}{r_2^2 s + s^2 x_2^2} \quad (9.1)$$

$s^2 x_2^2$ is neglected since it is very small

$$T = \frac{KSE_2^2}{r_2} = \frac{KSV^2}{r_2} \quad \text{or} \quad S = \frac{TR_2}{KV^2} \quad \text{or} \quad S \propto \frac{1}{V^2}$$

From the above expression, it can be seen that slip is inversely proportional to v^2. In other words, speed is directly proportional to v^2.

We know that torque is proportional to v^2. If the voltage is reduced by half, the torque gets reduced by 1/4 times. The circuit and waveforms are shown in Fig. 9.1.

The cheapest method of controlling the speed of an induction motor is to vary the applied voltage by using an A.C. chopper in each line. The speed can be decreased by increasing the firing angle and vice versa. In the three phase a.c chopper circuit, two SCRs conduct at a time. One SCR from the odd group and one from even group conducts. The SCRs conduct in the sequence 61, 12, 23, 34, 45, 56, ... To get this sequence, we have to number such that the difference of thyristor number in each anti parallel set is equal to 3.

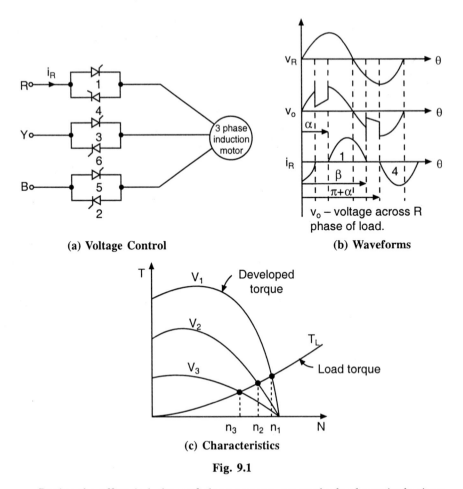

(a) Voltage Control
(b) Waveforms
(c) Characteristics

Fig. 9.1

During the off period, the emf alone appears across the load terminals since the current is zero.

Advantages 1. Smooth variation of speed 2. Energy saving is possible.

Disadvantages 1. Harmonics. 2. Electromagnetic interference.

9.2.2 V/f Control

The induction motor equivalent circuit is similar to that of transformer. We know that $E_1 = 4.44\ N\phi f$. If the stator impedance is neglected, V_1 is approximately equal to E_1. V/f ratio represents flux. If the voltage is kept constant and frequency is decreased, the flux might increase and the flux in machine may saturate. Thus the voltage and frequency have to be varied such that V/f ratio is kept constant. If voltage is increased, frequency also has to be increased and vice versa.

The variable voltage, variable frequency supply required for the speed control of induction motor is obtained by using a rectifier and inverter system as shown in Fig. 9.2.

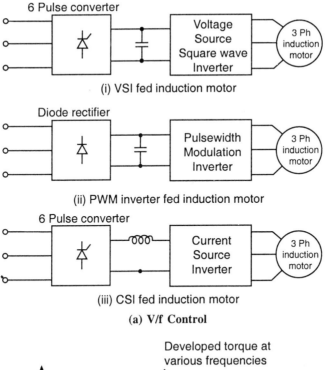

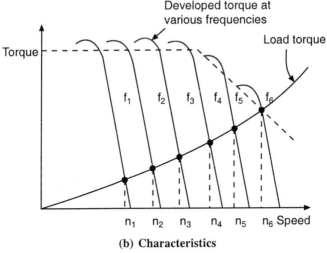

(b) **Characteristics**

Fig. 9.2

The variable voltage is obtained by using a fully controlled 6 pulse converter (Fig. 9.2(a)i). The rectifier and the filter capacitor act as a stiff voltage source. The square wave inverter converts the DC voltage into variable frequency AC voltage. The inverter can operate in 120° mode or 180° mode.

The diode rectifier converts fixed AC voltage into fixed DC voltage (Fig. 9.2(a)ii). The PWM inverter is capable of producing variable voltage and variable frequency supply. This supply is used to control the speed of induction motor.

In Fig. 9.2(a)iii, the full converter and DC link inductor act as a current source. The CSI converts fixed DC current into variable frequency AC current. The same is used to control the speed of induction motor.

The torque speed characteristic of V/f controlled induction motor drive are shown in Fig. 9.2b. Family of characteristics are shown for different frequencies. Base frequency (f_b) multiplied by K_1 is equal to rated voltage. Till the base frequency, the voltage and frequency are simultaneously increased. The flux is constant and therefore the torque is constant. The mechanical power increases linearly since the speed increases linearly ($p_m = T\omega$).

Beyond the base frequency, the voltage is kept constant at rated value and the frequency alone is increased to increase the speed. The flux decreases and therefore the torque decreases. But the speed increases since the frequency is increased. The mechanical power or H.P remains constant. Along with the motor characteristic, the load characteristic is also shown. Operating point is the point where the machine characteristic and the load characteristic intersect. If it operates at f_1, the corresponding speed is n_1.

9.2.3 Rotor Resistance Control

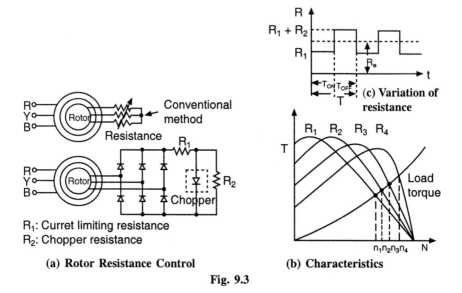

R_1: Curret limiting resistance
R_2: Chopper resistance

(a) Rotor Resistance Control   (b) Characteristics

Fig. 9.3

In conventional method, a ganged resistance is used to vary the resistance in the rotor circuit. The disadvantages of this system are sparking and maintenance.

A static variation in the rotor resistance can be obtained by using a chopper in the rotor circuit. The rotor DC chopper controlled induction motor drive is shown in Fig. 9.3. The variable frequency AC power in the rotor is converted to DC power using diode rectifier. R_1 is the current limiting resistance and R_2 the chopper resistance. Average value of resistance is the ratio of area to the base.

From Fig. 9.3c

$$R_e = \frac{R_1 T_{ON} + (R_1 + R_2) T_{OFF}}{T} = \frac{R_1 T_{ON} + R_1 T_{OFF} + R_2 T_{OFF}}{T}$$

$$= \frac{R_1(T) + R_2 T_{OFF}}{T} = R_1 + R_2(T_{OFF}/T) = R_1 + R_2 \frac{(T - T_{ON})}{T}$$

$$= R_1 + R_2(1 - T_{ON}/T) \tag{9.2}$$

$$R_e = R_1 + R_2(1 - \delta)$$

where δ is the duty ratio of chopper. By varying δ of the chopper, smooth variation of resistance and speed can be obtained. A family of characteristics are drawn for different values of duty cycle or different effective rotor resistance values. The speed can be increased from n_1 to n_4 by increasing the resistance from R_4 to R_1.

Advantages
1. Smooth variation of speed can be obtained.
2. It is easy to implement closed loop system using chopper.

Disadvantages
1. Firing and commutation circuits are required for the chopper.
2. The current on AC side of diode rectifier is a quasi square wave. This current injects harmonics into the rotor. This results in heating of the rotor.
3. Large P.I.V. rated SCR is required.

9.2.4 Slip Power Recovery Scheme

We know that $\dfrac{\text{Rotor output}}{\text{Rotor input}} = 1 - S$

$$\frac{P_o}{P_i} = (1 - S) \qquad P_o = P_i(1 - S) = P_i - SP_i$$

SP_i is called slip power.

The block digaram of slip power recovery scheme is shown in Fig. 9.4. In rotor resistance controlled drive, the efficiency is poor since the slip power is dissipated in the external resistance. The efficiency gets reduced at low speeds or high slip values. The slip power can be fed back to the AC source by using a rectifier inverter system. The rotor supply is a low voltage, low frequency supply. This is converted into DC by using an uncontrolled rectifier. The line commutated inverter converts the DC power into AC power. This AC power is fed to the mains through a transformer.

The circuit in Fig. 9.4 is called static krammer system. The fully controlled converter operates with α greater that 90° i.e., in the range 90-180°. The direction of the current cannot be reversed due to the presence of unidirectional devices (SCRs). The power can be made to flow from DC side to AC side by reversing the polarity of DC voltage.

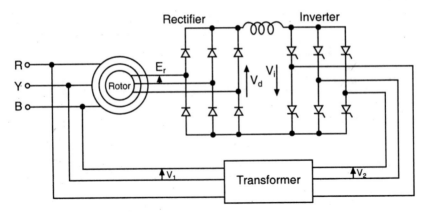

Fig. 9.4 Static Krammer System

Apply KVL in the D.C. loop by neglecting resistance of the filter inductance

$$V_d + V_i = 0; \quad V_d = -V_i$$

$$1.35\, E_{r(\text{Line to Line})} = -1.35\, V_1 \cos \alpha$$

$$E_r = S.E_{2(LL)} = -V_1 \cos \alpha; \quad S = \frac{-V}{E_{2(LL)}} \cos \alpha; \quad S = -a \cos \alpha \quad (9.3)$$

where $a = \dfrac{V_1}{E_2}$.

From eq. (9.3), it can be seen that slip and speed can be varied by varying firing angle α. This method is called speed control by emf injection method or static krammer system.

9.3 BRAKING OF INDUCTION MOTOR

The three types of braking are (1) Plugging, (2) Dynamic and (3) Regenerative braking.

9.3.1 Plugging

The circuit used for plugging is shown in Fig. 9.5. Under normal operating conditions, the antiparallel switches 1, 2, 3 are operating and the phase sequence of the motor voltages is *RYB*. Firing pulses to the switches 2A and 3A are suppressed. when braking is required, the firing pulses to 2 and 3 are stopped and firing pulses are given to 2A and 3A. Now the phase sequence of motor voltages is *RBY*. The direction of RMF gets reversed. Due to inertia, the speed cannot reverse instantaneously. The speed reduces to zero and if it is allowed, the motor runs in the opposite direction. To achieve braking, the supply to the induction motor is disconnected at the instant of zero speed. It is a fast method of braking.

Disadvantagess
1. Severe mechanical stresses on the rotor.
2. Requirement of SCRs with higher current rating.
3. High power loss.

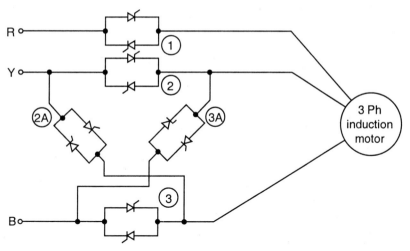

Fig. 9.5 Solid State Plugging

9.3.2 Dynamic Braking

Induction motor can be slowed down by applying DC voltage to the stator. To achieve DC dynamic braking, 3-phase AC supply is disconnected and DC voltage is applied to the stator. The force experienced by the rotor conductors due to this DC field produce the braking torque. This is similar to the braking torque in an energy meter. The other method of dynamic braking is rotor resistance control method. The speed can be reduced by increasing the resistance in the rotor. The circuit is shown in Fig. 9.6(a).

9.3.3 Regenerative Braking

(a) VSI Fed IM Drive

VSI fed IM drive has a voltage source inverter at the machine end and a six pulse converter at the source end as shown in Fig. 9.6(b). When the induction machine is motoring, source end converter operates as rectifier and machine end converter operates as an inverter.

With a single 6 pulse converter at the input side, regeneration is not possible since the electrolytic capacitor does not allow reversal of the voltage polarity and the SCRs do not allow the reversal of current. Regeneration is possible by

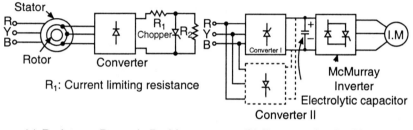

(a) Resistance Dynamic Braking   (b) Regenerative Braking

Fig. 9.6

having an antiparallel 6 pulse converter at the input side. When regeneration takes place, firing pulses to the converter 1 are stopped and firing pulses are given to machine end converter such that it operates as an inverter. The motor operates as an induction generator since it converts the mechanical power into electrical power. The diode rectifier part of VSI converts AC to DC. Converter II converts DC power into AC power and is fed to the AC mains. The disadvantage being that an additional six pulse converter is required for regeneration.

(b) CSI Fed Drive

The block diagram is shown in Fig. 9.7. This system has inherent regeneration facility. In the motoring operation, the supply end converter operates as rectifier and machine end converter operates as inverter. The power flows from the source to the machine. The induction machine acts as induction motor.

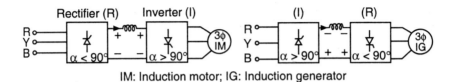

IM: Induction motor; IG: Induction generator

(a) Motor Operation   (b) Generator Operation

Fig. 9.7

In the regeneration mode, the machine end converter operates as rectifier and the source end converter operates as inverter. An external source drives the induction machine. It acts as induction generator. The machine end converter converts a.c power into dc power. The source end converter converts DC power into AC power. The power flows from machine to the a.c source. Hence it is called regeneration.

9.4 CLOSED LOOP OPERATION

Block diagram of closed system is shown in Fig. 9.8. Certain loads demand constant speed irrespective of the variations in the load. The disadvantage of open loop system is that the speed varies when the load varies. With closed loop system, a constant speed can be maintained. The closed loop system consists of two loops namely speed loop and current loop. The speed loop ensures that the motor runs at a speed equal to the set speed. The current loop ensures that the actual current is less than the rated current of the SCRs. Thus the current loop protects the SCRs from over currents. Voltage proportional to the actual speed can be obtained by using a tachogenerator (small dc generator). The speed error signal is the difference of set speed signal and actual speed signal. This is processed by the speed PI controller. The output of speed PI Controller is the set current since speed error is an indication of the load. The actual current can be obtained by using CT in the input side lines. The output of current comparator is $I^* - I$. This is processed through current PI controller. The control circuit vary the firing angle of the chopper using the output of

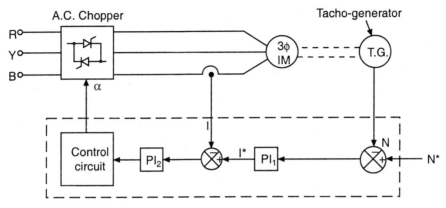

Fig. 9.8 Block Diagram of Closed Loop System

current PI Controller. If the load increases, the speed decreases. The slip increases and speed error increases and current error increases. The control circuit reduces the firing angle of a.c chopper. Thus speed of IM comes back to the previous value since the voltage is increased. Thus the closed loop system ensures that the actual speed is equal to set speed.

9.5 SPEED CONTROL OF DC MOTORS

The two basic methods of speed control are armature control and field control. In armature control method, field voltage is kept constant and armature voltage is varied. In field control, armature voltage is kept constant and field voltage is varied. Torque-speed characteristics with armature control and flux control are shown in Fig. 9.9(a). Speeds below normal speeds are obtained with armature control method. Speeds above normal speeds are obtained with field control. In armature control method, the mechanical power increases linearly with the increase in the speed ($P_m = T\omega$).

Speeds above normal speeds are obtained by reducing the flux. Torque gets reduced in the field control since the flux is reduced. Mechanical power remains constant since the speed increases at the same rate as the torque decreases. Armature control is preferred to field control since the armature inductance is less than field inductance. In other words, the response is faster with armature control than the field control.

9.5.1 One Quadrant dc Drive

The polarity of output voltage cannot be reversed since a semiconverter is used here. The current direction cannot be reversed since SCRs and diodes are unidirectional devices. Hence it is called one quadrant DC drive. The circuit of a semiconverter fed separately exited DC motor drive is shown in Fig. 9.9(b).

A transformer with two secondary windings is used. AC supply from one winding is given to the semiconverter. A.C from the other winding is given to the uncontrolled rectifier. The diode rectifier supplies constant dc voltage to the field. Variable dc voltage from the semiconverter is applied to the armature. The output voltage of the semiconverter can be varied by varying the

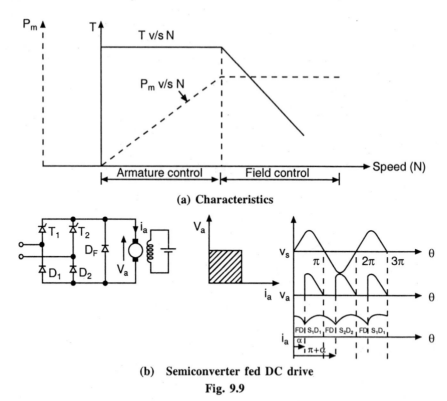

Fig. 9.9
(a) Characteristics
(b) Semiconverter fed DC drive

firing angle. Thus the speed of DC motor can be varied by varying the firing angle of the converter. The speed decreases with the increase in α and vice versa.

At α, T_1 is triggered. The devices T_1 and D_1 conduct from α to π. V_a and i_a are positive. Therefore power is fed to the motor. Beyond π, the current is transferred to D_F. During freewheeling v_a is zero. The armature current is maintained due to the conduction of D_F. D_F is forward biased by the voltage across armature inductance. Torque is developed by the armature during freewheeling due to the continuous current. During the negative half cycle, T_2 is triggered at $\pi + \alpha$. T_2 and D_2 conduct. Later the above process repeats.

9.5.2 Two Quadrant DC Drive

A full converter fed DC drive is shown in Fig. 9.10. The average Output voltage of this converter is positive with α between 0 to 90°. V_a is negative with α between 90–180°.

Two quadrant operation is possible. The first quadrant corresponds to motor operation and fourth quadrant corresponds to generator operation. In the process of generation, the mechanical energy drives the DC machine. The e.m.f will be more than the supply voltage. The DC machine acts as generator. The full converter acts as line commutated inverter. The DC power is converted into AC power. This AC power is fed to the mains.

Applications 125

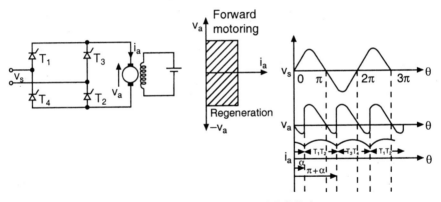

Fig. 9.10 Full Converter fed DC Drive

9.5.3 Four Quadrant DC Drive

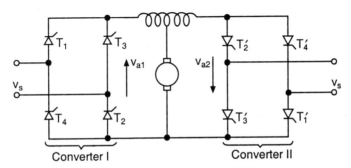

(a) Dual Converter Fed DC Drive

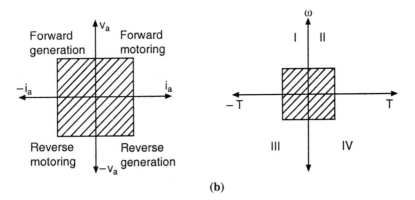

(b)

Fig. 9.11

Dual converter is shown in Fig. 9.11. If two full converters are connected back to back, it is called a dual converter. With converter *I*, first and fourth quadrant operations are possible. With converter *II*, second and third quadrant operations are possible since the current flows in a direction opposite to the reference current.

There are two modes of operation 1. Simultaneous control or circulating current mode 2. Non-simultaneous control or circulating current free mode.

In simultaneous control, converters operate such that $\alpha_1 + \alpha_2 = \pi$. A circulating current flows in the loop formed by the two converters since the instantaneous value of the output voltage of the two converters are not equal. The magnitude of circulating current is limited by a reactor.

In non simultaneous control, the firing pulses are given to only one converter at a time. Firing pulses are suppressed to the other converter. Change over from one the converter to the other converter takes a long time since the load is R-L-E load. Hence the response this method is slower than simultaneous control.

The four quadrant operations are as follows:

T (positive), ω (positive) is forward motoring
T (negative), ω (negative) is reverse motoring
T (negative), ω (positive) is forward regeneration
T (positive), ω (negative) is reverse regeneration

Since the product of T and ω is mechanical power.

9.6 BRAKING OF DC MOTORS

9.6.1 Plugging

The torque in a DC motor can be reversed by reversing the direction of armature current or flux. When the field connections are reversed, the polarity of back e.m.f. reverses. An external resistance has to be included to reduce the current. The disadvantages are severe mechanical stresses and power loss. The corresponding circuits are shown in Fig. 9.12(a).

9.6.2 Dynamic Braking

The supply to the armature is disconnected while the supply to the field is retained. The kinetic energy stored in the moving parts keep the rotor moving. The DC machine acts as separately exited DC generator since the field supplies the required flux. The kinetic energy is converted into electrical energy and the same is dissipated in the braking resistance. The braking torque can be contrtolled by varying the resistance R_b as shown in Fig. 9.12(b).

9.6.3 Regenerative Braking

The circuit of regenerative braking is shown in Fig. 9.12(c). Under normal operation the converter operates as rectifier. The contactors C_1 and C_2 get closed and the machine acts as a motor. For regeneration, the converter operates as inverter and the contactors C_3 and C_4 are closed.

9.7 CLOSED LOOP CONTROLLED DC DRIVE

Block diagarm of closed loop controlled DC drive is shown in Fig. 9.13. The closed loop system has two loops namely current loop and speed loop. The speed loop ensures that the drive runs at a speed equal to the set speed. The function of the current loop is to limit the current to a safe value. The closed loop system ensures that the drive runs at constant speed for fluctuations in the

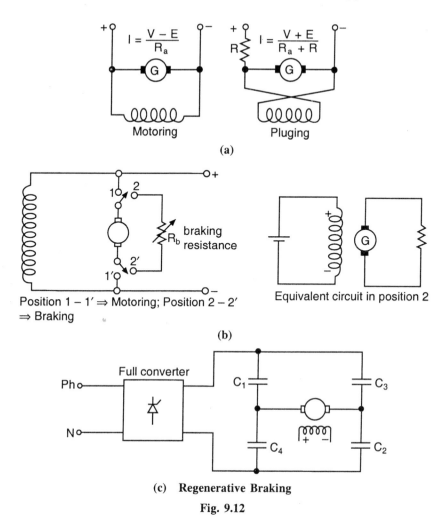

(a)

Position 1 – 1' ⇒ Motoring; Position 2 – 2' ⇒ Braking

(b)

(c) Regenerative Braking

Fig. 9.12

supply voltage and load. The actual speed can be measured by using a shaft encoder or tacho-generator. The output of shaft encoder is directly in digital form. If an analog tacho generator is used, ADC is required to convert the

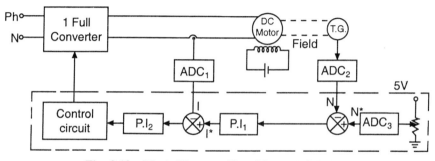

Fig. 9.13 Block Diagram Closed Loop DC Drive

analog signal into digital signal. The output of the counter gives actual speed. The speed error is processed by the speed PI controller. The current error is processed by current PI controller. The actual motor current can be sensed by using a shunt or a DCCT (DC current transformer). The analog signal proportional to the actual current is converted into digital signal since processor can handle only digital signals. If the load on the motor increases, the speed decreases. The PI controllers and the control circuit act such that the firing angle of the converter is reduced. (If it is a chopper controlled drive, the duty cycle is increased). The correction in the firing angle ensures that the motor runs at a speed equal to the set speed.

9.8a Regulated Power Supply (RPS)

D.C. regulated power supply systems are used in laboratories and electronic circuits of medical equipment etc. They are of two types (i) conventional regulated power supply (ii) switch mode power supply.

The circuit of conventional RPS is shown in Fig. 9.14. Commercial AC supply is reduced to a lower level by using a step-down transformer. Rectifier converts AC into unidirectional voltage which is a pulsating DC. This is filtered by using π-filter. The regulator converts unregulated DC voltage into constant dc voltage. The zener diode acts a voltage regulator..

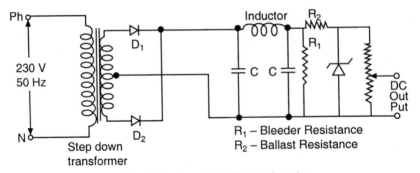

Fig. 9.14 Regulated Power Supply

Voltage across zener diode is applied to a potentiometer. Variable dc voltage can be taken from the output of potentiometer. A high resistance is connected in parallel with the filter capacitor. This is called bleeder resistor. The filter capacitor can discharge through the bleeder resistance when AC supply is disconnected. Under no-load condition, output current flows through the bleeder resistor and output voltage is available without load. Conventional power supply is bulky. This is due to the large size transformer and filter elements used in this circuit.

9.8b Switch Mode Power Supply (SMPS)

The disadvantages of the conventional regulated power supply unit are the requirement of a bulky transformer and bulky filter inductor. The size of transformer and filter inductor is reduced by using SMPS. The circuit of SMPS is shown in Fig. 9.15. The 230 V AC supply is directly given to a diode

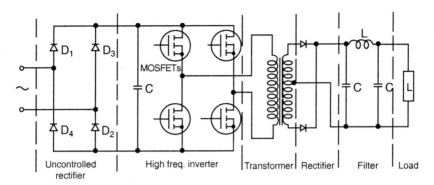

Fig. 9.15 Switch Mode Power Supply

rectifier. The DC output of the rectifier is the input to the high frequency inverter which operates at a switching frequency of about 200 KHz. The output of the inverter is reduced to a low voltage AC by using a step down transformer. This high frequency low voltage AC is rectified by using a centre tapped full wave uncontrolled rectifier. The output of the rectifier is filtered using L and C. The size of the filter components is reduced since the frequency of the ripple is high. From the emf equation of transformer, we know that emf depends on flux and frequency. If frequency of operation is increased, flux required gets reduced. Therefore, the size of the core gets reduced. The size of the transformer is very much reduced due to the high frequency operation. The transformer provides the isolation between the load circuit and ac input circuit.

9.9 WELDING

The process of fusing two pieces of metal together by passing heavy current through the area of contact where heat is required is called welding. The heat energy required for welding is given by $w = I^2 Rt$, where t is the time required for welding. The welding power is controlled by using AC chopper. Number of AC chopped pulses or integral number of AC cycles are applied using AC chopper. Current can be increased by using a step down transformer TR_1. The circuit and waveform with AC chopper are shown in Fig. 9.16.

In chopper control, the firing angle is decided by the time interval t. The disadvantages of phase control are low power factor, higher harmonics and EMI.

Integral cycle control can be used to control the welding power. In this control, voltage is applied to the load for few cycles (T_{on}) and voltage is not applied for period T_{off} by suppressing the pulses to the SCRs. The waveforms of integral cycle control are shown in Fig. 9.16(b) and 9.16(c).

The load is connected to the secondary of the transformer TR_2.

9.10 HEATING

Resistance Heating

Resistance heating may be achieved by using: (1) metal conductor of (2) non-metallic conductors e.g. carbon tubes and (3) liquids etc. Heating resistors are generally made of alloys of nickel, chromium and uranium. They are made in the form of wire or thin strips wound in the form of coils. Resistance heating is used in hot plates and heat insulated furnaces for laboratory and production purposes.

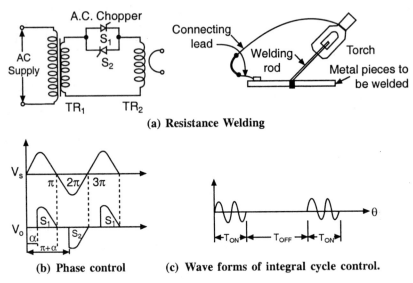

(a) Resistance Welding

(b) Phase control

(c) Wave forms of integral cycle control.

Fig. 9.16 Welding

Induction Heating

Induction heating is also known as eddy current heating because the heat produced is due to the eddy current losses taking place in the system. This type of heating is used in melting, annealing, forging, surface hardening, brazing and soldering operations. The principle of induction heating is as follows: The metallic job called charge is kept within the alternating field. When alternating voltage is applied to the job coil an alternating current starts flowing in the job coil. This current produces an alternating magnetic field. The metallic job placed in this field cuts the alternating magnetic flux and an emf is induced. The alternating current produced by the induced emf, is known as the eddy current. These eddy currents are responsible for generating the required amount of heat. As the supply frequency is increased the eddy current losses will increase which will cause more amount of heat to be produced.

Electronic Heaters Employed for Induction Heating Single-phase a.c. power is first rectified to produce direct current and is filtered to reduce the ripple content. This ripple-free d.c. is fed to an inverter circuit to produce high frequency power. The type of inverter used depends upon the value of frequency required for the heating process. High frequency power output of the inverter circuit is supplied to the job for heating it.

A non-conducting material generates heat when subjected to an alternating electric field. The process where in heating takes place due to dielectric loss is known as dielectric heating. The amount of heat produced depends on the value of the dielectric strength of the material. The non conducting material acts as the job which is to be heated. This method is extensively used in plastic and wood industries. It is especially of immense utility where multi plywoods are to be heated and glued. The heat supplied by this method is also employed in the textile, rubber, chemical and food industries. In this process, the job is placed in between few electrodes and the electrodes are fed with a very high frequency supply. The corresponding block diagram is shown in Fig. 9.17c.

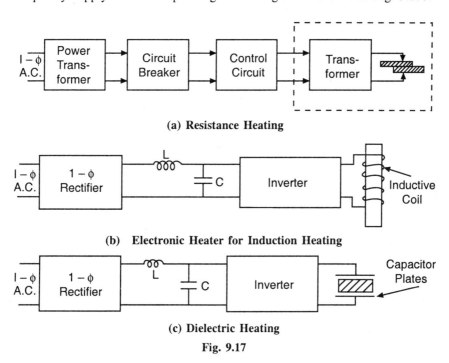

(a) Resistance Heating

(b) Electronic Heater for Induction Heating

(c) Dielectric Heating

Fig. 9.17

9.11 THYRISTOR CONTROLLED STATIC ON-LOAD TAP-CHANGING GEAR

A power system should maintain a constant voltage at the buses where loads are connected directly. Voltage variations are caused due to load fluctuations and because of the finite impedance of lines and cables. Voltage control is implemented by connecting transformers and by changing their turns ratio. A segment of HV winding of the transformers is usually provided with a large number of taps at different points. The turns ratio and therefore the output voltage can be changed with the help of a mechanical selector switch. The operation of a tap changer requires breaking and making of electrical contacts at taps. Most of the tap changer perform this operation at off-load. Tap-changing transformers are used where frequent voltage correction is not required. On load and where load disconnection is not accepted, the current is switched from tap to tap mechanically in the on-load tap-Characteristic (OLTC). This causes sparking, hence pitting and erosion of axing contacts. This contaminates the oil in the OLTC.

132 *Fundamentals of Power Electronics*

The other disadvantages of OLTC are its high cost, high maintenance, slow response and wide temporary voltage fluctuations.

Figure 9.18(a) shows the basic current of static control tap characteristic system. It consists of a three winding transformer (T_1) with the tertiary winding feeding a boosting transformer (T_2) through an antiparallel thyristor switch S_1. If the thyristor switch is triggered without delay (i.e. at the zero crossings of the current waveform), a constant voltage is added to the secondary voltage and in phase with it. A short circuiting antiparallel switch S_2 is required to prevent an open circuit condition of T_2 that occurs when the triggering of S_1 is delayed. The primary current of T_2 is dictated by the load current through the line but not by the secondary current as in a conventional transformer. The

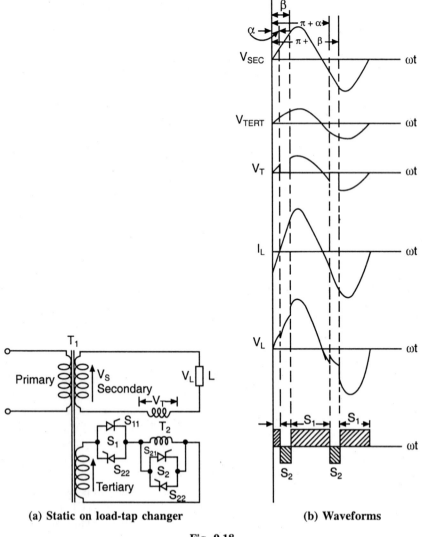

(a) **Static on load-tap changer**   (b) **Waveforms**

Fig. 9.18

secondary MMF opposes the primary MMF. If the booster transformer secondary is open circuited, the large primary MMF induce very high voltage in the secondary and large flux saturates the core. This is prevented by using the switch S_2.

The waveforms of the voltage and current are shown in Fig. 9.18(b) by assuming a lagging powerfactor. The delay angles α and β represent the delay of firing of the switches S_1 and S_2 respectively. The forward-biased thyristor of switch S_1 is S_{11}. S_{11} can be triggered at any time within the range $\theta < \alpha < 180°$, where θ is the phase angle difference between voltage and current waveforms. Similarly, the thyristors S_{21} and S_{22} can be triggered within the range $0 < \beta < \theta$.

The voltage across the tertiary winding of the transformer is a full sine wave. A segment of this voltage is available across the booster transformer from 0 to α by closing the switch S_1. Another segment of the tertiary voltage appears from β to $\pi + \alpha$ due to the conduction of S_1. These segments are added to the secondary voltage V_s to get V_L. Thus the booster transformer boosts the voltage at the load bus. The magnitude of boosting voltage can be controlled by controlling the firing angle delays of the thyristors.

ADVANTAGES
1. Continuous control of voltage is possible.
2. A fast acting closed loop system can be achieved.
3. No tap-changing gears are required.
4. Transient voltage dips are not there.

DISADVANTAGE
This scheme introduces harmonics into the load.

9.12 STATIC VAR COMPENSATORS
Presently thyristors are used for switching capacitors or inductors across the supply lines to compensate the load reactive power. Continuous and very fast control of reactive power has been made possible using thyristors.

The reactive power compensators are external devices which supply and compensate the lagging reactive power consumed by the load, thereby relieving the burden on the AC supply. The function of the static compensators is to minimise the voltage fluctuations at a given bus and to improve the supply power factor by compensating the load reactive power. The problem of compensation is viewed from two aspects, i.e. Load compensation and voltage support.

9.12.1 Load Compensation
The objective is to reduce or cancel the reactive power demand of large and fluctuating individual loads such as electric arc furnaces, rolling mills etc.

9.12.2 Voltage Support
This is generally related to the voltage at a given terminal of transmission line. The objective is to balance the voltages of the three phases, and control the voltage whenever it deviates from the reference value. The transmission voltage is regulated by the shunt compensators.

9.12.3 Shunt Reactive Power Compensators
1. Fixed capacitor banks with OCBs.
2. Thyristor switched capacitors (TSC).
3. Thyristor controlled Reactors (TCR).

(a) Fixed Capacitor Banks with OCBs
The leading current drawn by the shunt capacitors compensates the lagging current drawn by the load. They are installed across the supply lines at industrial service inputs. In the case of widely fluctuating loads, the VAR of the load also varies over wide limits. The fixed capacitor bank may often lead to either over-compensation or under compensation.

(b) Thyristor Switched Capacitors (TSC)
Depending on the total VAR requirement, a number of capacitors are used which can be switched into or out of the system individually. The control is done continuously by sensing the load VARs. If more compensation is required, then more capacitors are switched into the circuit and vice versa. The methods using circuit breakers and relays suffer from the drawback of being sluggish, unreliable, introducing switching transients and requiring frequent maintenance. With the advent of high power solid state devices, thyristors have replaced the mechanical switches. The rugged electronic circuits have made the system highly reliable. Thyristor switching has made it possible to achieve virtually continuous control of reactive power.

Each single phase thyristor switched capacitor branch consists of two major parts, the capacitor C and the thyristor switch as shown in Fig. 9.19(a). Each thyristor switch is built up from two thyristor stacks connected in antiparallel, one stack for each current direction. Each stack contains high-power thyristors in series, the number depending on the reated voltage of the switch.

Let us assume that capacitor voltage is not equal to the supply voltage when the thyristors are fired. Immediately after closing the switch, a large current flows and charges the capacitor to the supply voltage in a very short time. The SCR switch cannot withstand this stress due to high di/dt and would fail. In order to overcome this problem, a damping reactor is connected in series with the capacitor. This reactor keeps the di/dt within the capability of the thyristor.

The problem of achieving transient free switching of the capacitors is overcome by keeping the capacitors charged to the positive or negative peak value of the fundamental frequency network voltage at all times when they are in the stand by state. The switching-on instant is then selected at the time when the same polarity exists in the capacitor voltage. This ensures that switching on takes place at the natural zero passage of the capacitor current. The switching thus takes place with practically no transients. This is called zero current switching.

Switching off a capacitor is accomplished by suppresssion of firing pulses to the antiparallel thyristors so that the thyristors will block as soon as the current becomes zero. In principle, the capacitor will then remain charged to the positive or negative peak voltage and be prepared for a new transient free switching-on. This logic is called gate suppression logic.

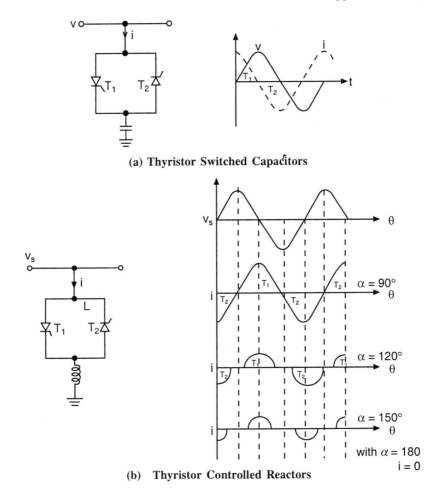

(a) Thyristor Switched Capacitors

(b) Thyristor Controlled Reactors

Fig. 9.19

(c) Thyristor Controlled Reactors

The basic principle of TCR is shown in Fig. 9.19(b). This shows one phase of a compensator. Two-way thyristor group is connected in series with a linear reactor. The thyristor group is built with two thyristor stacks in antiparallel, each stack containing a number of high power thyristors in series. By changing the firing angle, it is possible to change the value of the current continuously from zero to maximum value (Fig. 9.19(b)). Normally a step up transformer is used to suit the current ratings of the thyristors.

Partial conduction is obtained with higher values of firing angle delay. The effect of increasing the gating angle is to reduce the fundamental component of the current. This is equivalent to an increase in the inductance of the reactor, reducing its reactive power as well as its current. So far as the fundamental component of current is concerned, the TCR is a controllable susceptance, and can therefore be used as a static compensator. The current in this circuit is essentially reactive, lagging the voltage by 90° and this is continuously controlled by the phase control of the thyristors. The current is minimum at $\alpha = 180°$ and maximum at $\alpha = 90°$. Therefore the inductive reactance can be varied from minimum to maximum by varying α from $180°$ to $90°$.

Practically a combined TSC and TCR system is used for power factor control. This combination gives smooth variation of reactive power.

9.13 GENERREX EXCITATION SYSTEM OF ALTERNATORS

The generrex excitation system is a high initial response system with its power source as integral part of the generator itself. Excitation power is developed from the flux in the generator air gap (a three-phase voltage source) and from the current in the generator windings (a threephase current source). When the generator is operating at no-load, all of its excitation power is supplied by the voltage source. when the generator is operating with a load, excitation power is augmented by the current source. This system is self regulating with very high performance characteristics.

The outputs of potential source and of current source are combined in three excitation transformers to provide an ac output responsive to generator load, voltage, current and power factor. The potential source consists of a 3-phase water cooled winding, which is connected to the 'p' windings of the three/single phase excitation transformers and to three linear reactors. The potential source winding is located in the top of the generator stator winding slots. It is called generator excitation potential winding. The current source consists of three neutral leads of the generator stator winding which pass through the windows of the excitation transformers. This is called as 'C' winding. It is another input winding for the excitation transformer. This provides the additional field excitation needed under load conditions. The C winding also provides the excitation power during system faults when the system voltage and the machine air gap flux are at low level.

The output windings or 'F' windings of the excitation transformers are connected in delta and supply a three-phase bridge rectifier. The conversion from ac to dc is performed by rectifier consisting of silicon diodes. For control purposes, thyristors are connected as shown in Fig. 9.20. The output of the bridge rectifier supplies the current required by the generator field winding.

ADVANTAGES
1. Fast response.
2. Integral part of the synchronous machine.
3. Saves length of turbine generator block.
4. Diodes have lesser forward drop and they are more reliable than thyristors.

9.14 UNINTERRUPTIBLE POWER SUPPLY

Critical loads like computer systems, hospitals and air line reservation systems need uninterruptible power supply. A simple non-redundant UPS consists of a rectifier, a battery and an inverter as shown in Fig. 9.21(a). The rectifier converts the incoming power to dc which supplies the inverter. The inverter supplies constant voltage, constant frequency supply to the critical load. The battery is

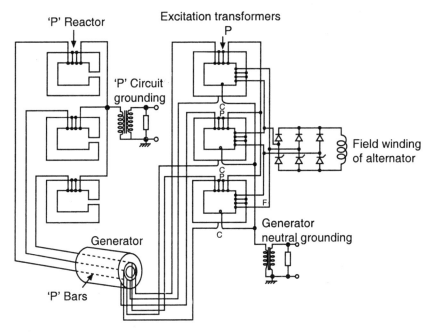

Fig. 9.20   Generrex Excitation System

floating on the DC bus. When there is a power outage, the battery provides the energy to the inverter. This system has the advantage of low cost and simplicity of operation. However, an internal failure of the inverter will cause a shut down. The battery and rectifier also have definite failure rates. Hence the system is not really uninterruptible.

9.14.1   Parallel Redundant UPS

The single line diagram of a parallel redundant system in shown in Fig. 9.21(b). The reliability can be increased by using more system components in parallel. The system mean time between failure (MTBF) as high as 80,000 hours can be achieved. The system consist of two parallel rectifiers each rated to carry full load, a battery and four parallel inverters rated such that full load can be supplied with two inverters shut down.

In the event of fault in one of the inverters, the fault has to be cleared instantly. The faulty inverter has to be isolated since it causes voltage dip on the critical bus. This is done by using the solid state interrupter.

Apart from continuity of supply, the UPS must also minimise voltage transients on the critical bus. The inverter must be designed to give a stable voltage during expected load changes. To improve the transient response, the inverter must have low inherent impedance.

The load and desired emergency operating time will decide the rating of the battery. It is possible to reduce the battery size and cost by using short time battery, until the diesel generator is brought up to feed power to the static UPS. This can assure power to the critical load even during extended power outages.

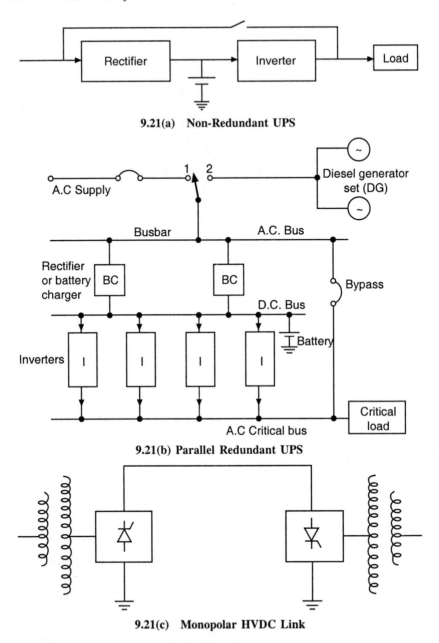

9.21(a) Non-Redundant UPS

9.21(b) Parallel Redundant UPS

9.21(c) Monopolar HVDC Link

A synchronized bypass is also provided so that the entire UPS can be removed from service without interrupting power to the critical load.

The control circuit ensures that all inverters operate in phase by controlling all inverters using one control signal. Hence any inverter can be taken in or out when the system is operating.

The DC bus is the isolating link between utility input disturbances, and the critical load bus. The battery can support the inverter system for nearly one

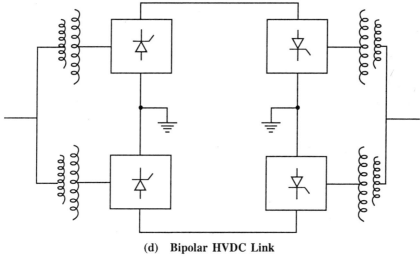

(d) Bipolar HVDC Link

Fig. 9.21

hour under full load. The diesel generators start automatically on loss of power to the rectifiers. A transfer switch then changes the UPS input bus from utility to the diesel set.

9.15 HVDC TRANSMISSION

HVDC can not be directly generated due to the limitations imposed by D.C. generator. Transformer can not be used to step up D.C. HVAC is converted into HVDC using a rectifier. At the receiving end, DC is converted into AC using an inverter. HVDC systems use monopolar or bipolar links. Monopolar system has only one conductor and earth is used as return conductor. Bipolar links have two conductors. At the sending end, the LVAC is stepped up using a transformer. The rectifier converts A.C. into D.C. The HVDC is transmitted over a long distance. The block diagram of HVDC system with monopolar link is shown in Fig. 9.21(c). Block diagram of bipolar link system is shown in Fig. 9.21d. This system has the following advantages:

1. Line construction is simpler.
2. Power transmission capability is higher since the reactance is zero.
3. No charging current since capacitive reactance is infinity.
4. No skin effect.
5. No compensation is required.
6. Less carona loss and less radio interference.
7. Higher operating voltages are possible.
8. There are no stability problems since the series reactance is zero.
9. Low short circuit currents.

It is economical to transmit power at high voltage and low current. Hence the HVDC transmission is preferred. The disadvantages are high initial cost and generation of harmonics.

9.16 MICROPROCESSOR BASED THYRISTOR CONTROLLED ELECTRICAL DRIVES

9.16.1 Speed Control of DC Motor Using Microprocessor

The AC input voltage is converted into variable dc voltage by using a fully controlled bridge rectifier. The variable dc voltage applied to the armature terminals is shown in Fig. 9.10.

The function of the firing circuit is to convert the firing angle input into a corresponding delay between the supply voltage and triggering pulse. The firing angle delay of thyristors is referred from the natural zero crossing which requires synchronization to a.c. supply. The interrupt handling capability of microprocessor provides an effective means to synchronize with supply voltage. The interface between microprocessor and supply is only a zero crossing detector (ZCD) and edge detector as shown in Fig. 9.22b.

The output of ZCD is a square wave with 180° ON and 180° OFF. The rising and falling edges of the square wave are detected using two monostable multivibrators. One is triggered at the rising edge of the input and the other triggered at the falling edge of the input. These outputs are OR'ed to get interrupt request (IR) signals. The IR signal is connected to RST 7.5 of the microprocessor. The corresponding waveforms are shown in Fig. 9.22a.

The controllable delay is provided by a programmable timer (8253), the operation of which is under the control of microprocessor. On receiving the IR signal, the processor reads alpha. The processor then calculates the count corresponding to this alpha and loads this count into the timer.

The timer is programmed to operate in mode 0 (interrupt on terminal count). Once mode-0 has been programmed, the timer out stays low. Now the counter starts decrementing the count value. When the count reaches zero, timer out goes to high level. To get a pulse, count register is reloaded with the mode.

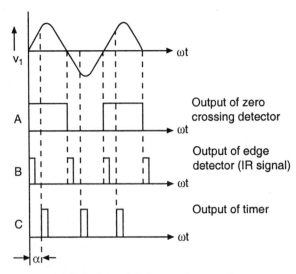

(a) Principle of firing angle control

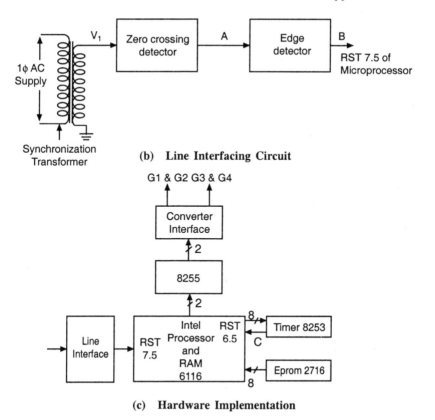

(b) Line Interfacing Circuit

(c) Hardware Implementation

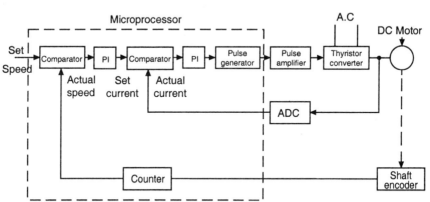

(d) Block diagram of microprocessor closed loop D.C. drive

Fig. 9.22

This timer output interrupts the processor through RST 6.5. The microprocessor distributes firing pulses of proper sequence.

The block diagram of microprocessor based closed loop dc drive is shown in Fig. 9.22(b). The firing angle of the converter is varied to obtain the speed control. The firing angle is varied using the difference between actual current and set current. The actual current is measured by using a DC CT. The set current is given by the difference between set speed and actual speed. The actual speed in digital system is measured by a shaft encoder. The output of the shaft encoder is fed to the microprocessor through I/O interface. The speed error and current error are computed by microprocessor. The PI controllers change their output as a function of the error. The firing angle of converter is varied such that the actual speed is equal to the set speed. To increase the speed of the motor, the firing angle is decreased, the average voltage applied to the motor is increased and the motor speeds up.

9.16.2 Speed Control of Induction Motor Using Microprocessor

The block diagram of closed loop induction motor drive system is shown in Fig. 9.23. The induction motor is fed from an AC to AC converter. Converter and inverter act as AC to AC converter. Converter is a 3-phase fully controlled bridge rectifier. Inverter can be a 3 phase current source inverter or a voltage source inverter. The speed of the induction motor is measured by using shaft encoder. The actual current on the ac input side is measured using a CT. This is rectified and fed to the ADC. The ADC will convert the analog current signal into a digital signal. The microprocessor with suitable software determines the speed error and current error. The current error modifies the firing angle of the rectifier. The speed error modifies the frequency of the inverter. The modifications are done such that the ratio v/f is constant. The microprocessor provides various control signals and acts as a control unit. This modifies the frequency such that the speed of induction motor is close to the set speed. The software consists of main routine and subroutines. In the main program, necessary initialization is done. The interrupts are enabled. The other routines are speed measuring routine, error generation routine, frequency correction routine, voltage correction routine and thyristor firing routine. These routines are executed when the processor is interrupted by the corresponding interrupt.

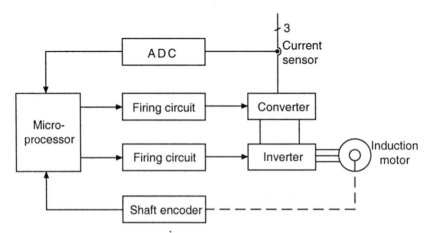

Fig. 9.23 Microprocessor Based Closed Loop Induction Motor Drive.

9.16.3 Microprocessor Based Synchronous Motor Drive

The block diagram for the speed control of synchronous motor is shown in Fig. 9.24. The microprocessor used has the following functions.

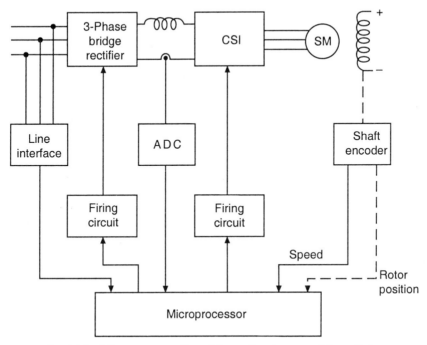

Fig. 9.24 Microprocessor Based Loop Synchronous Motor Drive

1. It has to ensure commutation of the inverter during starting and low speeds.
2. An automatic change over must occur from forced commutation to natural commutation when speed of the drive is 10% of rated speed.
3. Proper distribution of firing pulses to the rectifier and inverter.

The variable voltage and variable frequency supply is made available to the motor through a rectifier and inverter system. The disadvantages of open loop control are: (1) hunting, (2) problem of instability and (3) poor dynamic behaviour. If the motor uses closed loop self control, the above disadvantages can be eliminated. The firing pulses for the inverter are derived from a rotor position sensor in order to maintain absolute synchronisation between rotor and stator rotating magnetic field. The stator frequency is slaved to the rotor speed. Instead of speed following the frequency as in the case of conventional operation, the frequency follows the speed in self control. The angle between armature MMF and rotor MMF can be controlled using self control. This eliminates the instability problem. The shaft encoder shown in the figure provides the information of both speed and rotor position. The 3-phase fully controlled bridge rectifier and the dc link inductor acts as the current source.

The various routines for current regulation and speed regulation are executed

144 Fundamentals of Power Electronics

by means of proper interrupt signals generated by timers and line interfacing circuit.

List of Formulae for dc motor drives

$$T = K\phi I_a \tag{9.4}$$

$$E_b = K\phi\omega \tag{9.5}$$

$$T = \left(0.159 \frac{ZP}{A}\right)\phi I_a; \quad E_b = \frac{\phi ZN}{60} * \frac{P}{A}; \quad \omega = \frac{2\pi N}{60}$$

In separately excited machine, flux is constant. The constant $K\phi$ is called torque or emf constant.

From (9.5)

$$\omega = E_b/K\phi; \quad \omega = \frac{V_0 - I_a R_a}{K\phi} \tag{9.6}$$

(1) If DC Motor is fed from chopper $V_0 = \delta E$

(2) If DC motor is fed from 1ϕ full converter $V_0 = \dfrac{2V_m}{\pi} \cos\alpha$

(3) If it is fed from semi converter $V_0 = \dfrac{V_m}{\pi}(1 + \cos\alpha)$

(4) If it is fed from 3ϕ full converter $V_0 = 1.35\, V_1 \cos\alpha$

Problems

Ex. 9.1 The speed of a 12 H.P, 230 V separately excited DC motor is controlled by a single phase full converter. The rated current of the motor is 40 A. $R_a = 0.2\,\Omega$ and AC supply voltage is 260 V. The motor voltage constant $K\phi = 1.74$ volt-sec/rad. For continuous current conduction, Calculate motor torque and speed for a firing angle of 30°

Solution

$K\phi = 1.74$ Volt-sec/rad; $I_a = 40$ A; $R_a = 0.2\,\Omega$; $V_{rms} = 260$ V; $\alpha = 30°$

Therefore, torque $(T) = K\phi . I_a = 1.74 * 40 = 69.6$ N-m

$$\omega = \frac{E_b}{K\phi} = \frac{V_0 - I_a R_a}{K\phi}$$

but $$V_0 = \frac{2V_m}{\pi}\cos\alpha = \frac{2*260*\sqrt{2}}{\pi} = 202.8\text{ V}$$

$$\frac{2\pi N}{60} = \frac{202.8 - 40*0.2}{1.74}; \quad N = 1070 \text{ RPM}$$

Ex. 9.2 The speed of 125 HP, 600 V separately excited DC motor is controlled by a 3ϕ full converter which is operated form a 3ϕ, 480 V, 60 Hz supply. The rated armature current is 165 A. $R_a = 0.0874\,\Omega$, $K\phi = 0.33$ V/r.p.m. Calculate the full load speeds at $\alpha = 0$ and $\alpha = 30°$. Assume that the no load or armature current is 10% of the rated current and continuous.

Solution

$$K\phi = 0.33 \text{ V/r.p.m.} = 0.33 * \frac{60}{2\pi} \text{ V-sec/rad} = 3.15$$

At $\alpha = 0$, $I_0 = 10\%$ of rated current or 10% of 165 = 16.5 A

$$\omega = \frac{V_0 - I_a R_a}{K\phi}; \quad V_0 = 1.35 \, V_1 = 1.35 * 480 = 648 \text{ volt}$$

$$\omega = \frac{648 - 16.5 * 0.0874}{3.15}$$

Therefore $N = 1960$ r.p.m.
At $\alpha = 30°$

$$I_a = 16.5 \text{A}$$

But $\quad V_0 = 1.35 * V_1 \cos \alpha = 1.35 * 480 \cos 30° = 561.8$ volt

Therefore $\quad \omega = \dfrac{V_0 - I_a R_a}{K\phi}; \quad \omega = \dfrac{561.8 - 16.5 * 0.087}{3.15}; \quad N = 1696$ r.p.m.

Ex. 9.3 Determine the effective resistance of slip ring induction motor controlled by a DC chopper system. The chopper resistance is 2 Ω. A resistance of 4 Ω is connected in series with the chopper. $T_{off} = 4$ μsec and chopper frequency is 200 Hz.
Solution

$$T = \frac{1}{f} = \frac{1}{200} = 5 \, \mu \text{ sec.} \quad T_{on} = T - T_{OFF} = 5 - 4 = 1 \, \mu\text{sec}$$

$$\delta = \frac{T_{on}}{T} = \frac{1}{5} = 0.2 \quad R_e = R_1 + R_2(1 - \delta) = 4 + 2(1 - 0.2) = 5.6 \, \Omega$$

Ex. 9.4 A 3φ, 4 pole, 50 Hz slip ring induction motor is controlled by chopper TRC system (time ratio control system) $R_1 = 4 \, \Omega$, $R_2 = 2 \, \Omega$, $T_{OFF} = 4$ μsec. Choppper frequency is 200 Hz. For a motor slip of 2%. Calculate the torque developed if the DC current of the chopper is 10 A.
Solution

$$T = \frac{1}{f} = \frac{1}{200} = 5 \, \mu \text{ sec} \quad T_{on} = T - T_{OFF} = 5 - 4 = 1 \, \mu\text{sec}$$

$$\delta = \frac{T_{on}}{T} = \frac{1}{5} = 0.2 \quad R_e = 4 + 2(1 - 0.2) = 5.6 \, \Omega$$

$$T = \frac{I_d^2 R_2}{S \omega_S}; \quad N_s = 120 f/P = \frac{120 * 50}{4} = 1500$$

$$\omega_s = \frac{2\pi N_s}{60} = \frac{2\pi * 1500}{60} = 157 \text{ r/sec} \quad T = \frac{10^{2} * 5.6}{0.02 * 157} = 165 \text{ N-m}$$

Ex. 9.5 A 3φ, 60 Hz, 8 pole wound rotor induction motor is connected to 4160 line. The open circuit rotor line voltage is 1800 volt. If the motor has to develop 800 kw at a speed of 700 r.p.m, Calculate

(a) Slip Power
(b) D.C. Line Voltage
(c) D.C. Line Current

The speed of the motor is controlled by slip power recovery scheme.

Solution

$$f = 60 \text{ Hz}, \ P = 8, \ E_{12} = 1800 \text{ V}, \ P_0 = 800, \ N = 700 \text{ r.p.m.}$$

$$N_s = \frac{120 * 60}{8} = 900; \quad S = \frac{900 - 750}{900} = 0.222$$

$P_i: P_0: P_{cu} = 1 : (1 - S) : S$

$$\frac{P_{cu}}{P_0} = \frac{S}{1-S} \qquad P_{cu} = \text{slip power}$$

$$P_{cu} = \left(\frac{S}{1-S}\right) P_0 = \frac{0.22}{1-0.222} * 800 = 228 \text{ kw}$$

$$V_d = 1.35 \ E_r = 1.35 \ SE_2 = 1.35 * 0.222 * 1800 = 540 \text{ volt}$$

Slip Power = $V_d I_d$

$$228 * 1000 = 540 \ I_d; \qquad I_d = 422 \text{ A}$$

Short Questions and Answers

1. For speed control of D.C. motor using rectifiers, armature voltage control gives _____

    Ans. Below rated speed

2. Speed of induction motor can be controlled by _____

    Ans. Variable voltage and variable frequency

3. PWM control is suitable for _____

    Ans. Constant torque and constant power

4. HVDC transmission with converters is economical _____

    Ans. Above 800 Km

5. Voltage controlled induction motor operating at 20% slip will have a limiting efficiency of _____

    Ans. 80%

6. RF dielectric heating is used for _____

    Ans. Heating plastics

7. State and explain the merits and demerits of CSI over VSI?

    Ans. Merits: (1) Regeneration is possible with CSI
    (2) Simple power circuit

    Demerits: (1) Large filter inductor is required in the D.C. line
    (2) Power factor of the source end converter is poor at higher values of α.

8. What are the disturbances on the power lines which can harm the sensitive equipment like a computer?

    Ans. (1) Voltage fluctuations (2) Frequency fluctuations
    (3) Switching and lightning surges

10

Resonant Inverters

10.1 STATIC POWER CONVERSION

Switch mode dc to ac inverters are used in ac drives and uninterruptible power supplies where the objective is to produce a sinusoidal ac output whose magnitude and frequency can be controlled. The voltage at the terminals of ac motor must be adjustable in its magnitude and frequency. This is accomplished by means of the switch mode dc to ac inverter which needs a dc voltage at the input and produces the desired ac output voltage.

In recent years, the state of art of micro/macro power electronic conversion has advanced. Hence the systems are required to have high power density, fast transient response, high system stability, reduced EMI, less harmonic line current and dc ripple current. The above requirements necessitate that the power electronic conversion must be done effectively at higher switching frequencies.

Pulse width modulation (PWM) technique is commonly employed in inverters feeding induction/synchronous motors, so as to minimise torque pulsations. But in the PWM inverter, the increased switching losses, due to the simultaneous existence of non-zero device voltage and current during switching period reduces the efficiency of power conversion.

In resonant inverters, the switching losses are almost eliminated as the switching on and off of the device is possible when the device voltage or current is passing through zero. The resonant dc link inverter has numerous applications at high power levels because it does not demand large peak current or peak voltage switching devices unlike other types of resonant inverters.

10.2 ADVANTAGES OF ZERO VOLTAGE SWITCHING

1. Low switching losses and switching stresses in switching devices by turning them on and off at zero voltage and current.
2. Further reduction in switching stresses and losses in switching devices by generating only sinusoidal voltages and currents.
3. Higher switching frequencies operation.
4. Less Electromagnetic interference.
5. Reduced size magnetic components.
6. No acoustic noise problem.
7. Higher kilowatt power levels.

8. Considerable size and weight reductions.
9. Higher efficiency.

10.3 DISADVANTAGES OF ZERO VOLTAGE SWITCHING

In the zero voltage switching topology, the switch is required to withstand a forward voltage that is higher than V_d by an amount $Z_o I_o$ where Z_o is the characteristic impedance of the circuit. For zero voltage turn off of the switch, the load current I_o must be greater than V_d/Z_o. Therefore if the output load current I_o varies in a wide range, then the foregoing two conditions result in a very large voltage rating of the switch. Therefore this technique is limited to an essentially constant load application.

10.4 RESONANT DC LINK CONVERTER

10.4.1 Introduction

In the conventional switch mode PWM dc-ac inverters, the dc input voltage to the inverter is fixed and the output voltage is obtained by switch mode PWM switching. However in DC link converter, the dc bus is made to oscillate at high frequency as the bus voltage goes through periodic zero crossings, thus setting up ideal switching conditions for all the devices connected across the bus. Integral pulse modulated resonant inverter (IPMRI) can also be represented by this converter in certain modes of operation.

10.4.2 Operating Principle

Consider the circuit shown in Fig. 10.1 where dc voltage is used to supply energy to the load. A resonant LC circuit is inserted between the supply and load. A power transistor is connected across the capacitor C to control the output voltage. The power transistor is driven on and off at a fixed frequency f_s and at a low duty cycle as shown in Fig. 10.2. When the transistor T_1 is turned on for a period 't_{on}', the current through the inductor L rises linearly. During the turn off period 't_{off}' of the transistor, a sinusoidal current flows

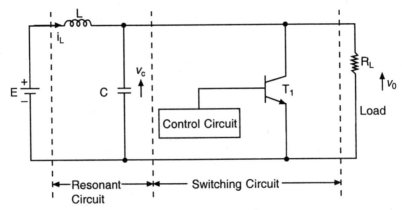

Fig. 10.1 Resonant DC Link Converter

through the capacitor C due to resonance. In the first half period of quasi-sine wave, the capacitor is charged to twice the dc input voltage and then it is discharged to zero during the next half period. This process of switching on and off at a fixed frequency with low duty cycle is repeated to obtain quasi-sinusoidal voltage across the capacitor.

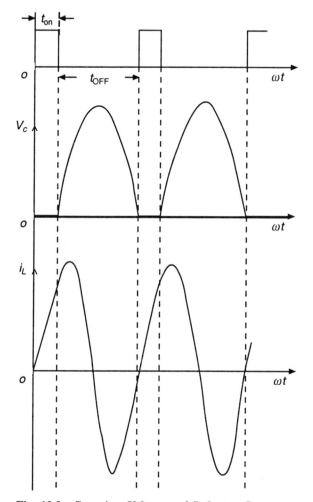

Fig. 10.2 Capacitor Voltage and Inductor Current

The base drive circuit of the power transistor used in the resonant dc link converted is explained in the next section.

10.5 BASE DRIVE CIRCUIT OF BIPOLAR JUNCTION TRANSISTOR

The base drive for bipolar juction transistor (BJT) needs to do the following functions. It must turn the transistor on in the shortest time to minimise the turn on losses. It must provide sufficient steady state base current to keep the device saturated. It is required to turn the transistor off as quickly as possible when the drive is removed.

10.5.1 Design Consideration

The design of base drive circuit for power transistor is considerably more complicated than the design of gating circuit of thyristors. The required base currents will be large because of the low beta value of the power transistor and consequently logic circuit cannot drive power transistors. An intermediate gain stage made up of transistors of moderate power and current capability must be used to provide the large base current needed to drive the high power transistor. This leads to significant power dissipation in the driver circuit and in the main power transistor which has to be considered.

10.5.2 Description of Base Driver Board

The base driver circuit shown in Fig. 10.3 is designed to achieve the above mentioned design considerations. The circuit demands a 12-0-12 V regulated power supply. The regulated power supply is obtained from the circuit consisting of a 15-0-15 V centre tapped transformer, bridge rectifier circuit with filter and series regulators 7812 and 7912.

The base drive is obtained from the 8255 PPI. Opto coupler (MCT2E) is used to isolate the microprocessor and the power circuit. The anode terminal of the opto coupler is connected to the 8255 PPI of the microprocessor and the cathode to the ground of the microprocessor system.

Whenever the anode terminal of MCT2E goes high, it will turn on transistors Q_1, Q_4, Q_3 and Q_5 respectively developing a positive voltage across the 150 ohms resistor. This voltage is applied between base and emitter terminals of the power transistor. When the anode terminal of MCT2E is low, then Q_1, Q_4, Q_3, and Q_5 are turned off, Q_2 is turned on and hence the voltage across the 150 ohms resistor provides reverse bias to the base of power transistor to turn it off quickly

10.6 SINGLE PHASE IPMRI

10.6.1 Introduction

Most of the developments for ac drives has been to improve the output voltage and current waveforms and to reduce the switching losses in the power switches. The voltage source inverter configuration has simple power structure and control strategy is resonably simple. This configuration also provides a fully regenerative interface between the dc source and the ac load. However, the following problems are identified with this configuration.

- Low switching frequencies as a result of high switching losses.
- High dv/dt at the output generates interference due to capacitive coupling.
- Diode reverse recovery and snubber interactions cause high device stress under regeneration condition.
- Poor fault recovery characteristics.

Resonant Inverters **151**

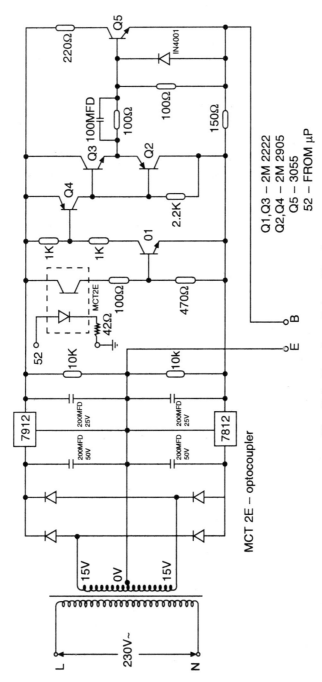

Fig. 10.3 Transistor Driver Circuit

152 *Fundamentals of Power Electronics*

- Acoustic noise at the inverter switching frequency can be very objectionable.

However, increase in inverter switching frequency with minimised lower order harmonics is achieved with PWM inverter, but the switching losses are high. Hence an optimum power converter with zero switching losses, high frequency switching capability, multi quadrant operation capability and small reactive components is to designed. This can be achieved with the quasi resonant inverter where the dc bus is made to oscillate at a high frequency so that the bus voltage goes through periodic zero crossings, thus setting up ideal switching conditions for the devices connected across the bus.

10.6.2 Single Phase IPMRI with R-Load

The single phase transistorised IPMRI shown in Fig. 10.4 converts the dc voltage to a low frequency ac voltage consisting of an integral number of pulses of high frequency. The resonant circuit consisting of L_2 and C_2 is provided between the dc source and the transistor bridge circuit.

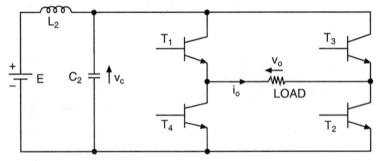

Fig. 10.4 Single phase IPMRI

The operation of IPMRI with R-Load can be divided into three modes as shown in Fig. 10.5. The waveforms of the output voltage with four pulses per half cycle and base drive pulses to various transistors of IPMRI system are shown in Fig. 10.6.

Mode I (t_0, t_1)

The current through the resonant inductor increases linearly when T_1 and T_4 are turned on simultaneously for a period t_{on}. The voltage across the capacitor is zero. This mode ends at the instant t_1, when the base drive pulse for T_4 is inhibited.

Mode II (t_1, t_2)

A positive resonant pulse of the output voltage is initiated by driving T_2 on. The source supplies both the load current and the capacitor charging current. Now the stored energy of the inductor is transferred to the capacitor. This mode ends when the capacitor voltage reaches 2E.

Resonant Inverters 153

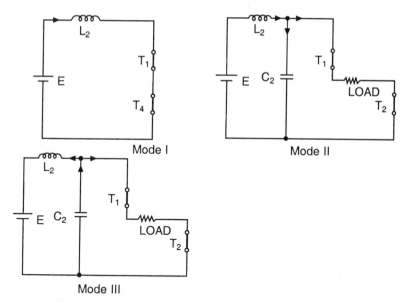

Fig. 10.5 Modes of Single Phase IPMRI with R-Load

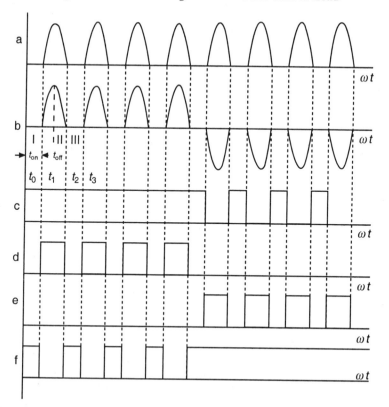

Fig. 10.6 Waveform with R-Load: (a) Capacitor voltage (b) Output voltage (c–f) Base drive pulses to $T_1 - T_4$

Mode III (t_2, t_3)
The base drive pulses to transistors T_1 and T_2 are retained. The capacitor discharges through the load and the dc source. This mode ends when the capacitor voltage reaches zero.

This process of turning on and off is repeated 'p' times during every half period of the output voltage. Thus a low frequency ac voltage consisting of an integral number of high frequency pulses is generated. Hence this inverter is called high frequency integral pulse modulated resonant inverter.

10.7 ANALYSIS OF IPMRI WITH R-LOAD
When two transistors T_1 and T_2 or T_3 and T_4 are conducting, the load resistance R_L appears in parallel with the resonant capacitor. The circuit configuration for this condition reduces to resonant dc link converter circuit.

The quasi-sinusoidal voltage at the resonant dc link is generated by proper selection of t_{on} and the same is obtained by equating the energy stored in L_2 during t_{on} and the energy dissipated in R_L during t_{off}. Let the energy stored during t_{on} and t_{off} be W_{on} and W_{off} respectively.

$$W_{on} = \int_0^{t_{on}} V(t)i(t)\,dt = \frac{V^2}{2L_2} t_{on}^2 \qquad (10.1)$$

$$W_{off} = \int_0^{t_{off}} \frac{V^2}{R_L}(1 - \cos \omega t)^2\,dt = \frac{1.5\,V^2}{R_L} t_{off} \qquad (10.2)$$

Equations (10.1) and (10.2) must be equal for producing quasi-sinusoidal voltage, therefore

$$t_{on} = \frac{\sqrt{3L_2\, t_{off}}}{\sqrt{R_L}} \qquad (10.3)$$

where
$$t_{off} = 2\pi\sqrt{L_2 C_2} \qquad (10.4)$$

From equation (10.3) it is seen that the energy to be stored in the inductor during t_{on} depends on the load resistance and resonant frequency.

10.8 SINGLE PHASE IPMRI WITH R-L LOAD
The circuit operation of IPMRI with R-L load can be divided into six modes as shown in Fig. 10.7. The associated waveforms with two pulses per half cycle are shown in Fig. 10.8.

Mode I (t_0, t_1)
The current through the resonant inductor increases linearly when T_1 and T_4 are turned on simultaneously for a period t_{on}. The voltage across the capacitor is zero. This mode ends at instant t_1, when the base drive pulse for T_4 is inhibited.

Mode II (t_1, t_2)
A positive resonant pulse of the output voltage is initiated by driving T_2 on.

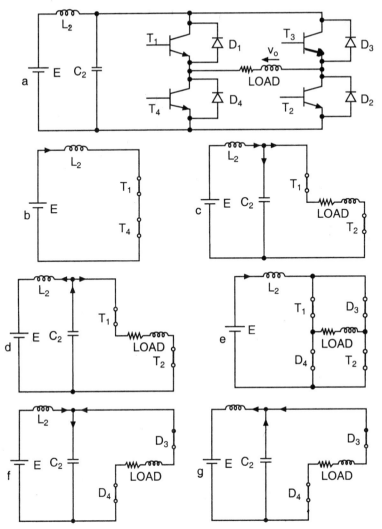

Fig. 10.7 Modes of single phase IPMRI with R-L Load: (a) Inverter Circuit (b–g). Modes I to VI

The source supplies both the load current and the capacitor charging current. Now the stored energy of the inductor is transferred to the capacitor. This mode ends when the capacitor voltage reaches 2E.

Mode III (t_2, t_3)
The base drive pulses to transistors T_1 and T_2 are retained. The capacitor discharges through the load and the dc source. This mode ends when the capacitor voltage reaches zero.

Mode IV (t_3, t_4)
The diodes D_3 and D_4 are forward biased by the load voltage. A small overlap

156 *Fundamentals of Power Electronics*

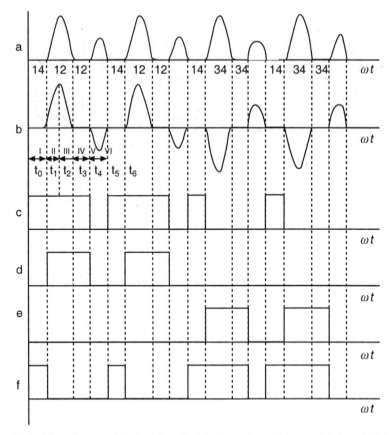

Fig. 10.8 Waveforms with *R-L* Load: (a) Capacitor Voltage (b) Load Voltage (c-f) Base Drive Pulses to T_1–T_4

exists since T_1 and T_2 do not turn off instantaneously. There are three current paths. The first path is through load, D_3 and T_1, the second path through load, T_2 and D_4 and the third path through D_4, T_1, L_2 and dc source. The voltage across the capacitor remains zero during this mode.

Mode V (t_4, t_5)
The base drive pulses for all the transistors are inhibited. The energy stored in the load inductor is fed back to the capacitor. The current drawn by the capacitor is the sum of the source current and the load current. This mode continues till the capacitor voltage reaches its peak value.

Mode VI (t_5, t_6)
Diodes D_3 and D_4 remain forward biased. The current through the capacitor reverses and the current through the dc source is the sum of capa-citor and load current. This mode ends when the capacitor voltage reaches zero.

The above modes are repeated 'p' times per half cycle of the output frequency for obtaining integral pulse modulated voltage at the output. During the negative half cycle of the output voltage, the transistors T_3 and T_4 conduct and the circuit operation is similar to the preceding half cycle.

11

Quasi Resonant Converters

11.1 GENERAL

In switching converter circuits, magnetic elements and capacitors play the major roles of energy storage and transfer and ripple filtering. Since the required values and volumes of magnetic elements and capacitors decrease as the operating frequency is increased, it is imperative to design converter circuits capable of operating at high frequencies to achieve high power density. However, switching an inductive load at high frequencies as in the case of switching converters, imposes high switching stresses and switching losses on semiconductor devices. The conventional approach to relieve the devices of their stresses has been the use of snubber circuits. A typical snubber circuit is as shown in Fig. 11.1. It is important to note that by using a snubber network, the switching losses and stresses are merely transferred and not eliminated.

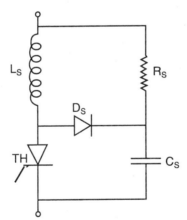

Fig. 11.1 Typical Snubber Circuit

11.2 ZERO CURRENT SWITCHING

A fundamental departure from the conventional "forced turn-off" approach is the "Zero Current Switching" (ZCS) technique, proposed by F.C.Y. Lee et al. Replacing the power switches in the PWM converters by resonant switches gives rise to a new family of converters, namely "quasi resonant converters". This new family of converters, can be viewed as a hybrid between PWM converters and resonant converters. They utilize the principle of inductive or capacitive energy storage and transfer in a similar fashion as PWM converters. The circuit topologies also resemble those of PWM converters. However, an LC tank circuit is always present near the power switch and is used not only to shape the current and voltage waveforms of the power switch but also to store and transfer energy from input to output in a manner similar to the conventional resonant converters.

11.3 RESONANT SWITCH TOPOLOGIES

A resonant switch is a sub-circuit consisting of a semiconductor switch and its associated LC resonant elements for waveform shaping. The sinusoidal current waveform generated by the LC resonant circuit creates a zero current condition for the switch to turn off without switching stresses and losses. The actual implementation of the resonant switch can be either in a "half wave" configuration where the current can flow only in the forward direction as in Fig. 11.3(b) or in a "full wave" configuration, where the current can flow bidirectionally as in Fig. 11.3(c). There are two types of resonant switch configurations, L-type and M-type. In both cases L_1 is connected in series with S_1 to limit the di/dt of the switch current, and C_1 is added as an auxiliary energy storage/transfer element. L_1 and C_1 constitute a series resonant circuit with its oscillation initiated by the turn on of S_1.

If the ideal switch S_1 is implemented in a unidirectional configuration, as shown in Fig. 11.3(b), the resonant switch is confined to operate in a half wave mode. If diode D_1 is connected in antiparallel with TH, as shown in Fig. 11.3(c), then the resonant switch operates in a full-wave mode and the switch current can flow bidirectionally.

In essence, the LC resonant circuit is used to shape the current waveform through switch S_1. At turn on, the device voltage (V_{CE} or V_{DS}) can be driven into saturation before the current gradually rises in a quasi-sinusoidal fashion. Because of the resonance between L_1 and C_1, current through S_1 will oscillate to a negative value, thus allowing it to be naturally commutated. Fig. 11.2 gives the locus of the voltage and current with and without the snubber circuits.

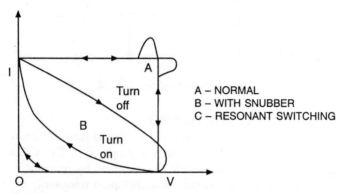

Fig. 11.2 Locus of Switch Voltage and Current During Switching

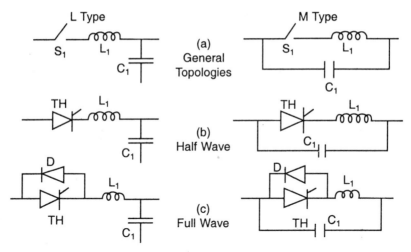

Fig. 11.3 Resonant Switch Configurations

As it can be seen, the voltage and current ratings have to higher if hard switching is performed. The safe operating region is also greater in the case of resonant switching.

11.4 PRINCIPLE OF OPERATION OF QRC

A conventional zero current switching quasi-resonant converter in half wave mode is shown in Fig. 11.4(a). Its corresponding full wave mode is shown in Fig. 11.6(a). To analyse its steady-state behaviour, the following assumptions are made:-

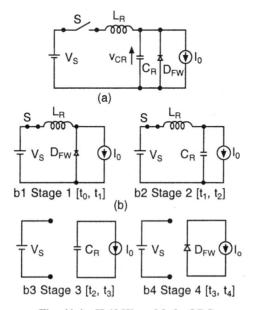

Fig. 11.4 Half Wave Mode QRC

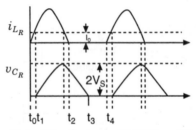

Fig. 11.5 Waveforms in Half Wave Mode

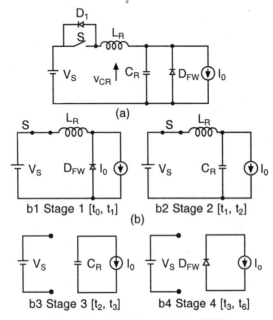

Fig. 11.6 Full Wave Mode QRC

(i) Filter inductance is much larger than resonant inductance.
(ii) Output filter and load are treated together as a constant current sink.
(iii) Semiconductor switches are ideal (i.e.) no forward voltage drops in the on-state, no leakage currents in the off-state and no time delays during both turn-on and turn-off.
(iv) Reactive elements of the tank circuit are ideal.

The following variables are defined

$$\text{Characteristic impedance } (Z_n) = \sqrt{L_R/C_R} \qquad (11.1)$$

$$\text{Resonant angular frequency } (\omega) = 1/\sqrt{L_R C_R} \qquad (11.2)$$

$$\text{Resonant frequency } (f_r) = \omega/2\pi \qquad (11.3)$$

A switching cycle can be divided into four stages. The associated equivalent circuits for these stages are shown in Fig. 11.4 and Fig. 11.6 for half wave and full wave modes respectively.

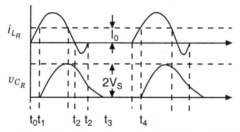

Fig. 11.7 Waveforms in Full Wave Mode

Assume that initially diode D_{FW} carries the output current I_o and resonant capacitor voltage V_{CR} is clamped at zero and switch S is off. At the beginning of the switching cycle $t = t_0$, S is switched on.

11.4.1 Inductor Charging Stage (t_0, t_1)

When S is turned on at $t = t_0$, the input current i_{LR} rises linearly and is governed by the state equation

$$L_R \left[\frac{di_{LR}}{dt} \right] = V_s \qquad (11.4)$$

The duration of this stage, td_1 ($t_1 - t_0$) can be solved with boundary condition $i_{LR}(0) = 0$ and $i_{LR}(td_1) = I_o$.

$$\text{Thus } td_1 = \frac{L_R I_o}{V_s} \qquad (11.5)$$

11.4.2 Resonant Stage (t_1, t_2)

At time $t = t_1$ when the input current rises to the level of I_o, D_{FW} is turned off and the amount of current ($i_{LR}(t) - I_o$) is now charging C_R. The state equations are

$$C_R(dV_{CR}/dt) = i_{LR}(t) - I_o \qquad (11.6)$$

$$L_R(di_{LR}/dt) = V_s - V_{CR}(t) \qquad (11.7)$$

with the initial conditions

$$V_{CR}(0) = 0 \text{ and } i_{LR}(0) = I_o$$

Therefore

$$i_{LR}(t) = I_o + (V_a/Z_n) \sin \omega t \qquad (11.8)$$

$$V_{CR}(t) = V_s (1 - \cos \omega t) \qquad (11.9)$$

If a half wave resonant switch is used, switch S will be naturally commutated when the resonating input current $i_{LR}(t)$ reduces to zero. On the other hand, if a full wave resonant switch is used, current $i_{LR}(t)$ will continue to oscillate and energy is fed back to source V_s through D. Current through D again oscillates to zero. The duration of this stage td_2 ($t_2 - t_1$) can be solved by setting $i_{LR}(td_2) = 0$.

Thus,
$$td_2 = \alpha/\omega \tag{11.10}$$

where $\alpha = \arcsin\left[\dfrac{Z_n I_o}{V_s}\right]$

$\pi \leq \alpha \leq 3\pi/2$ for half wave mode
$3\pi/2 \leq \alpha \leq 2\pi$ for full wave mode
At time t_2, V_{CR} can be solved using
$$V_{CR}(td_2) = V_s(1 - \cos \alpha) \tag{11.11}$$

11.4.3 Capacitor Discharging Stage (t_2, t_3)

This stage begins at t_2, when the current through the inductor L_R is zero. At $t = t_2$, S is turned off. Capacitor C_R discharges through the load to supply constant load current. Hence V_{CR} decreases linearly and reduces to zero at t_3. The state equation during this interval is

$$C_R(dV_{CR}/dt) = I_o \tag{11.12}$$

The duration of this stage $td_3(t_3 - t_2)$ can be solved with the initial condition

$$V_{CR}(0) = V_s(1 - \cos \alpha) \tag{11.13}$$

$$td_3 = C_R V_s(1 - \cos \alpha)/I_o \tag{11.14}$$

11.4.4 Free-wheeling Stage (t_3, t_4)

This stage starts with conduction of freewheeling diode and the load current freewheels through D_{FW} for a period td_4 until T_1 is turned on again. The duration of this stage is $td_4(t_4 - t_3)$.

12

Microprocessor Based Triggering Schemes

The advantages of microprocessors are flexibility, universal hardware, low cost, fault diagnostic ability, easy interfacing and fast operation.

The disadvantages are lack of access to intermediate signals, difficult design of digital filters and large assembly language program development time.

Two types of microprocessor based triggering schemes for 3-phase converters are discussed in this chapter.

12.1 FIRING SCHEME FOR 3 PHASE CONVERTERS PROPOSED BY HUY, ROYE AND PERRET

This scheme is based on 8085 processor along with 8253 timer and 8255 programmable peripheral interface. The synchronisation of the firing signals with the line is obtained by means of a line interface which consists of a zero crossing detector and an edge separator. The output of this edge separator goes as an input to the gate '0' of 8253. Totally two counters of 8253 are used, i.e., counter '0' and counter '1'. The connection diagram is shown below.

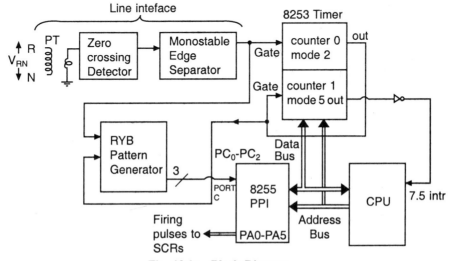

Fig. 12.1a Block Diagram

164 *Fundamentals of Power Electronics*

The counter '0' is programmed in mode 2 and loaded with a count corresponding to 60°. Counter '1' is programmed in mode 5 and loaded with a count corresponding to the firing angle (α if α is < 60°, else if α > 60° it is loaded with α-60, else if α > 120°, it is loaded with α-120°). So the displacement between out '0' and out '1' will be corresponding to the firing angle α or α-60 or α-120.

At any instant, the decision regarding the thyristors to be fired is taken based on 3 factors (i) It is taken only when the interupt occurs (ii) the decision is based on the line status of R, Y and B phases (iii) It is also dependent upon the range which the firing angle lies i.e., whether it lies between 0 and 60° or 0 and 120° or 120 and 180°. So three look-up tables are derived in accordance with the *RYB* pattern and the three ranges of firing angles.

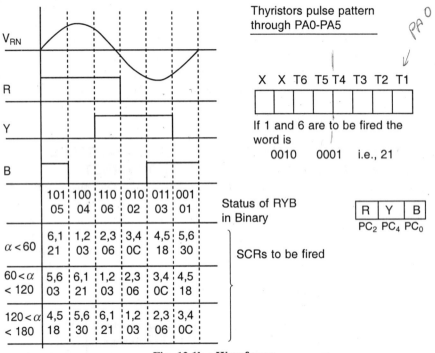

Fig. 12.1b Waveforms

The three look-up tables are:

0° < α < 60°		60° < α < 120°		120° < α < 180°	
RYB	Firing pattern	RYB	Firing pattern	RYB	Firing pattern
00	00	00	00	00	00
01	30	01	18	01	0C
02	0C	02	06	02	03
03	18	03	0C	03	06
04	03	04	21	04	30
05	21	05	30	05	18
06	06	06	03	06	21

Once these tables are stored in the memory, the only thing pending is to write an appropriate software. The *RYB* pattern generator can be designed as a sequential circuit. The flow chart and firing diagrams are given below:

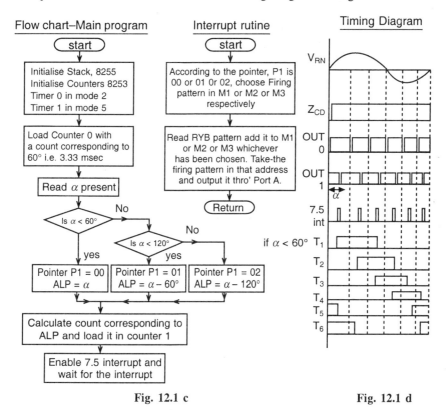

Fig. 12.1 c   Fig. 12.1 d

Disadvantages of this firing scheme

(1) Frequency deviations are not taken into account. If frequency value is not exactly 50 Hz and if it varies, the 60° count that has been loaded may be incorrect due to which the pulses to thyristors may not be given properly.

(2) When the firing angle is decreasing, let us say 62° to 58° one interrupt pulse will be missing. This has to be probably taken care of in the software (i.e. inside the interrupt routine).

12.2 FIRING SCHEME PROPOSED BY S.B. DEWAN

This scheme is also based on 8085 along with 8253 timer. Two Counters of 8253 are used — counter '0' for taking care of any supply frequency derivations and counter 1 for firing the thyristors. The Counter '0' is programmed in mode '0' and counter 1 in mode 2. The hardware diagram of the scheme is shown in fig. 12.2a. The working of the scheme is as follows: the zero crossing of V_{RN} is detected and the square wave is given to the gate of counter '0' of 8253 which is programmed in mode 0. At every positive going zero crossing of V_{RN}, the processor is interrupted at 6.5 level and a count of FFFF is loaded into the counter '0' inside the interrupt routine.

166 *Fundamentals of Power Electronics*

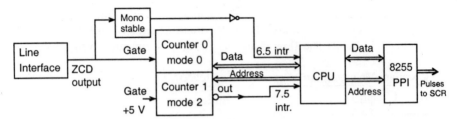

Fig. 12.2a Block Diagram

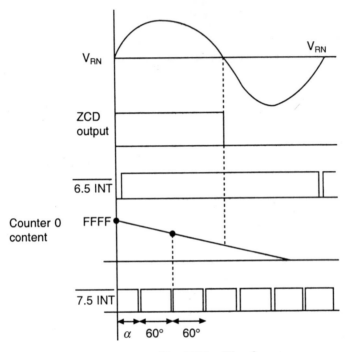

Fig. 12.2b Waveforms

The counter starts counting down from the moment its gate goes high. At any instant the counter '0' contents 0 let us say the counter '0' contents and after 90° from the zero crossing. Then if the frequency of V_{RN} has been 50 Hz, the counter '0' contents should be $\left[FFFF - \frac{90}{360} \times 1/50 \times \frac{1}{T_{CLK}} \right]$ i.e., for the 5 m sec in how many times the counter '0' contents would have been counted is given by [5 msec/CLK where CLK is the clock of the 8253. If the frequency is not exactly 50 Hz, the contents of counter '0' would be different from that of [FFFF − 5 × 10⁻³ × T_{CLK}]. Difference between this actual and ideal values gives the frequncy deviation.

So, everytime the counter '0' contents are read at the 2nd 7.5 interrupt (let us say so). If $\alpha = 30°$, the 2nd interrupt will occur exactly at 90°. So depending on α we can calculate the contents of counter 0 register at the time of reading (rather, this is the predicted value). When the counter is read, **the difference** between the actual and predicted values give the error in frequency which can be divided by 6 and this error can be distributed to the 60° count so that it will be taken care of by the 8253 timer.

Counter 1 is initially loaded with α count, so the first interrupt pulse will be $\alpha°$ displaced from the zero crossing. After the counter 1 gives the pulse for the first time at its OUT terminal, it is later on loaded with 60° count. Now the counter will continuously produce pulses at 60° intervals as it is programmed in mode 2. At each consecutive pulse the thyristor pulses are given appropriately (which may be deduced from the firing pattern that was outputted in the previous interrupt).

If the firing angle is to be changed, let us say from 65° to 64° then, this difference of 1° is adjusted in the following manner. Just for one interval, counter 1 is loaded with 59° (instead of 60°) so that the interrupt pulse is produced 1° ahead. Now this firing will be done at an angle equivalent to 64°. After this, the count in counter 1 is again retrieved to 60° itself so that this new firing angle of 64° is maintained throughout the forthcoming cycles. The flow charts for the various routines are shown below:

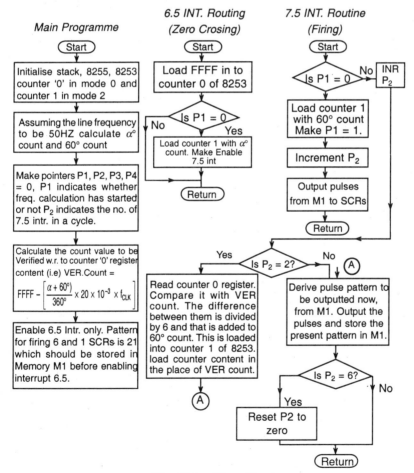

Fig. 12.2c Flow Chart

Appendix 1
Basic Experiments in Power Electronics

In this appendix, the following basic experiments are discussed:

1. SCR characteristics
2. UJT firing circuit
3. Triac characteristics
4. R-C triggering scheme of SCR
5. Voltage commutation
6. Current commutation

1. SCR CHARACTERISTICS

AIM
To obtain the following of a given SCR.

(i) $V_g - I_g$ characteristic,
(ii) $V_{FB} - I_g$ characteristic.

The above characteristics can be obtained by using the experimental set up shown in Fig. A1.1.

(I) $V_g - I_g$ Characteristic (Gate characteristic)

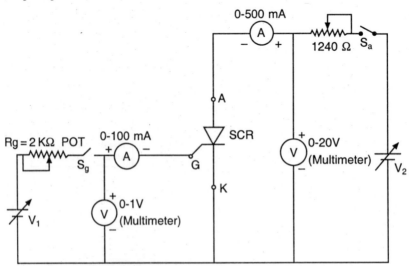

Fig. A1.1   Circuit Diagram to Experimentally Determine SCR characteristics

Theory

Silicon Controlled Rectifier (SCR) is a four layer PNPN device. It has three terminals. The outermost P layer is the anode, the outermost N layer is the cathode and the inner P layer is the gate. This device acts as a controlled switch. The switch is made to close by triggering it under forward biased condition.

There are various methods of triggering the SCR and among them the most popular one is the gate triggering method. When a positive voltage is applied to the gate with respect to the cathode, keeping anode also positive referred to the cathode, a current flows into the gate and this current triggers the SCR. The relationship between the gate voltage V_g and the gate current I_g is known as the gate characteristic. There is a PN junction in between the gate and cathode and hence the gate characteristic will be similar to a diode characteristic. This characteristic is useful in designing the trigger circuit.

Procedure

It is better to do this part of experiment by connecting V_1, voltmeter, ammeter and pot on gate side.

1. Keep the switch S_g open.
2. Adjust V_2 to about 5 V.
3. Keep R_g at its maximum value.
4. Close the switch S_g alone.
5. Note down the gate voltage V_g and the gate current I_g.
6. Decrease R_g and for each value of R_g, note down V_g and I_g.

The values of V_g and I_g must be within their maximum values specified in the data sheet.

Observation

V_g						
I_g						

Graph

Draw a graph between the gate voltage V_g and the gate current, I_g.

(II) $V_{FB} - I_g$ Characteristic (Blocking characteristic)

Procedure

1. Keep S_a and S_g open and have about 5 V in V_2 and 20 V in V_1.
2. Adjust R_c to such a value that the current in the SCR will be more than the latching current when the SCR conducts.
3. Adjust R_g to its maximum.
4. Close S_a and S_g and slowly increase I_g by decreasing R_g until SCR conducts.

5. Note down minimum gate current I_g minimum and the forward voltage V_{FB} across SCR before it conducts.
6. Open both S_a and S_g and increase V_1 in steps and repeat steps 2 to 5.

Observation

V_{PB}							
I_g							

Graph
Draw a graph between the minimum gate current I_g and the voltage V_{FB}.

2. UJT FIRING CIRCUIT

AIM
To study and design UJT oscillator circuits for a given frequency and varying the trigger delay angle.

Theory
The unijunction transistor (UJT) is a three terminal device with only one junction. It is of n-type silicon wafer with a p-type alloy junction. Two leads are connected to the ends of the n-material and they are called as bases B_1 and B_2. The lead connected to p material is called as emitter. When the emitter junction is not forward biased, the UJT behaves as high resistance. This resistance is called interbase resistance R_{BB}.

The resistance R_{BB} can be considered as the series connection of two resistances R_{B1} and R_{B2}. R_{B1} is the resistance between the junction point and the end of base B_1 within the wafer and R_{B2} is between the junction point and the end of base B_2.

At 25°C the value of R_{BB} lies between 4 and 10 kohm and it increases linearly with temperature.

In the normal operation, B_1 is connected to ground of the supply and B_2 is connected to positive end of the supply. They are connected either directly or through resistors. The voltage between B_1 and B_2 is called as the interbase voltage V_{BB}. The interbase voltage allows the leakage current to flow through R_{BB} causing a voltage drop across R_{B1}. The ratio of this voltage to the interbase voltage V_{BB} is called as the intrinsic stand—off ratio. The value of η lies between 0.51 and 0.82 and it is about 0.63 for the popularly used UJT (2N2646).

The emitter junction forms a diode when the voltage applied to the emitter is less than the cut-in voltage of diode. The emitter junction is reverse biased and only the leakage current flows through the UJT. As the emitter voltage (V) is increased, the diode gets forward biased at a particular emitter voltage and a large emitter current flows through B_1. This emitter voltage is called peak point potential (Vp). This voltage can be expressed as $Vp = \eta\, V_{BB} + V_D$ where

V_D is the diode voltage of nearly 0.5 V. Once the diode conducts it remains in conduction even after V_E is reduced below V_p. But below a certain lower voltage the diode again stops conduction and this lower voltage at emitter is called as the valley voltage (V_v).

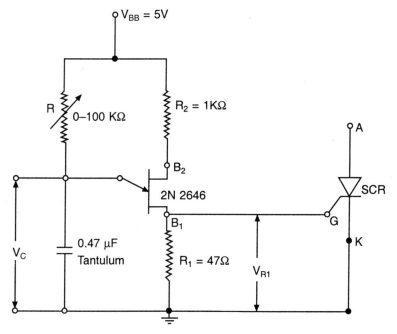

Fig. A1.2 Circuit Diagram for UJT Firing Scheme

Procedure
Do the connections as per the circuit shown above. Measure practical time period and compare it with theoretical value. Practical time period can be obtained using CRO. Theoretical time can be calculated using the formula $T = RC \ln 1/(1 - \eta)$. Observe the pulses across R_1 using CRO.

3. TRIAC CHARACTERISTICS

AIM
To obtain the characteristics of the given triac.

Components Required
Triac ST44; Rheostat 50 Ω, 210 Ω; Ammeter 0–500 mA; RPS; Digital voltmeter.

Theory
Triac is a bidirectional controlled switch. It is used extensively for AC phase control. The three terminals of a triac are the main terminals 1 and 2 and the gate. A triac can be triggered by a positive or negative gate pulse.

172 Fundamentals of Power Electronics

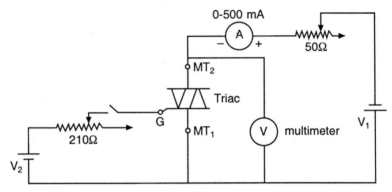

Fig. A1.3 Circuit Diagram for Triac Characteristics

With no signal at the gate, the triac will block both the half cycles of the A.C applied voltage in case the peak voltage is less than the break over voltage of the triac. The triac can be turned ON in each half-cycle by applying a suitable gate pulse. The sensitivity of the triac is high when both MT2 and gate are positive or negative whereas the sensitivity becomes less when MT2 and gate are of opposite polarity.

In general the modes of operation of the triac are

$MT_2 + Ve \quad MT_1 - Ve \quad G + Ve \quad OR - Ve \rightarrow$ I Quadrant operation

$MT_2 - Ve \quad MT_1 + Ve \quad G + Ve \quad OR - Ve \rightarrow$ III Quadrant operation.

Procedure
Do the connections as shown in the figure. Make the terminal MT2 positive with respect to MT1. Turn on the triac by making gate positive with respect to MT1. Note down the voltage across the triac for various load currents by varying the resistance in the main circuit.

Make gate negative with respect to MT1 and repeat the above process. Plot the V-I graph in first and third quadrants.

4. RC TRIGGERING SCHEME OF SCR

AIM
To trigger the given SCR at various firing angles by a half wave RC triggering scheme.

Components Required
Potentiometer 0-100 kΩ; capacitor 0.47 µf; resistors 1.5 kΩ, 33 kΩ, 1 kΩ; diode IN4001; SCR; Lamp load.

Theory
SCR is a controlled switch which is made to close by triggering it under forward biased condition. Gate triggering is the desirable method of turning on SCR. When the SCR is in the forward blocking state. It can be turned on by

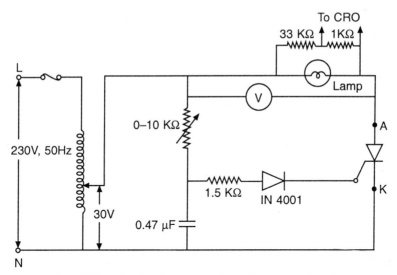

Fig. A.1.4 Circuit Diagram for RC Halfwave Triggering

applying a pulse at the gate, which is positive with respect to the cathode. In half wave RC triggering, the voltage across a capacitor is used to trigger the SCR. A negative voltage should not be applied to the gate of the SCR when it is in the reverse blocking state. This is ensured by having a diode D in the gate circuit. The diode conducts only when the capacitor voltage is positive i.e. during the positive half cycle. So phase control of only a half cycle is possible using this arrangement. The firing angle can be varied by varying the RC time constant of the circuit. The SCR is naturally commutated at zero crossing of applied voltage.

Procedure

1. Do the connections according to circuit diagram.
2. Adjust the output of the auto transformer to 30 volts. The brightness of the lamp load can be varied by adjusting the resistance of the potentiometer.
3. Observe voltage waveform across the lamp load.
4. Note the conduction time and delay time for different settings of the potentiometer and calculate firing angles from this.

5. VOLTAGE COMMUTATION

AIM
To study the effect of initial voltage across the commutating capacitor and the value of capacitor on the commutating currents.

Theory
Commutation is a process by which the current from one SCR is transferred

174 *Fundamentals of Power Electronics*

to another SCR or a conducting SCR is turned off. To have a successful commutation, three conditions have to be satisfied and they are

(a) The current flowing through the SCR is brought to zero.
(b) A reverse voltage V_{rev} is applied across the SCR, and
(c) The reverse voltage is maintained across the SCR for a period more than the turn off time t_{off} of the SCR.

Unless all the above three conditions are met with, the SCR cannot be successfully commutated. The above conditions are easily met with if the voltage applied to the SCR is alternating or the load connected to the SCR is under damped. The process of commutation of former type is called as input or natural commutation and the later is called as the load commutation. When the input voltage is a dc voltage, some extra circuit is to be used to commutate the SCR. The commutation of this type is called as forced commutation. The forced commutation circuits generally employ a charged capacitor with proper polarity of voltage across it. The charged capacitor is brought either directly across the SCR under commutation or through an inductor. In voltage commutation, a charged capacitor is directly connected across the SCR and in current commutation a charged capacitor is connected across the SCR through an inductance. In the former case the capacitor voltage directly helps in commutation whereas in the latter the building up of the current in the LC circuit helps in commutating the SCR.

Referring to Fig. A1.5. When the charged capacitor is connected across the conducting SCR, the capacitor discharges a current through the SCR and the current is brought to zero almost instantaneously (condition a). After the SCR current has become zero, the capacitor charges through R. The capacitor voltage starts decreasing and it attains zero value after sometime. Until the voltage becomes zero the capacitor applies a reverse voltage across the SCR (condition b). The reverse bias time depends upon the value of the capacitor, the initial

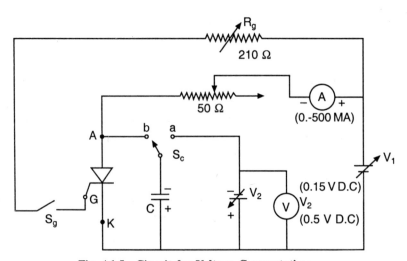

Fig. A1.5 Circuit for Voltage Commutation

voltage across the capacitor and the current through the SCR at the start of the commutation period. The value of the capacitor and its initial voltage have to be properly selected to have reverse bias period more than the turn off time of the SCR (condition c). The maximum current that can be commutated for a given capacitor and initial voltage V_i is called as the maximum commutating current I_{com}.

Procedure

Adjust V_1 to about 10 V and have maximum value in R_L such that the current through the SCR will be about 0.5 A when the SCR conducts.

1. Adjust V_2 to 5 V
2. Keep S_C in position 'a'.
3. Trigger SCR by closing S_g and adjusting R_g. After triggering *SCR*, open S_g.
4. Throw S_C to position 'b' and check whether the SCR is turned off or not.
5. If the SCR is turned off, open S_c. Trigger SCR again
6. Reduce R_L to increase the current in SCR.
7. Repeat steps 3 to 7 until the commutation fails. Note the maximum anode current that is commutated and the voltage in V_3 (i.e.) V_i.
8. Repeat steps 3 to 7 for different initial voltages V_i.
9. Repeat steps 1 to 8 for different values of capacitor.

Graph

Plot the following

1. Maximum commutating current I_{com} versus the initial voltage V_i across the capacitor for different capacitor values.

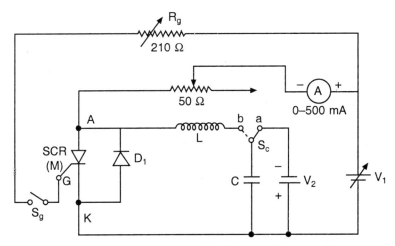

Fig. A1.6a Circuit Diagram for Current Commutation

6. CURRENT COMMUTATION

AIM
To study the effect of the initial voltage across the commutating capacitor on the commutating currents.

Theory
In the experiment on voltage commutation, it has been explained how the voltage across the commutating capacitor helps to commutate the SCR. In the other class of commutation called as current commutation, a sinusoidally rising current is used to commutate the SCR. The principle of this class of commutation can be understood by considering the following example.

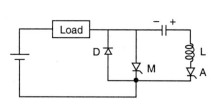

Fig. A1.6b   Basic circuit   Fig. A1.6c   Current waveform

Let the capacitor C be initially charged to some voltage with the polarity as shown in Fig. A1.6b.

The triggering of the main SCR M allows load current I_L to flow through the load.

To commutate M, the auxiliary SCR A is triggered. Since A is forward biased, it conducts and forms a ringing circuit with C, L and M. The capacitor current increases sinusoidally. During this period, two currents I_L and I_C flow through M and they are in opposite directions. As the capacitor current increases the net current in M decreases when $ic = I_L$, the current in M becomes zero, and M is turned off. After M is turned off the capacitor current continues to flow through D.

The current i_c reaches a peak value and then decreases. When it becomes equal to I_L again (during decreasing period) D is turned off.

When D is conducting, it applies a reverse voltage across the commutated SCR M. For successful commutation this reverse bias period should be greater than the turn off time of M. Once M and D are turned off, the load current flows through C, L and A and the load current gradually reduces to zero.

Procedure
Adjust V_1 to about 10 V and have the maximum value of R_L shown in Fig. A1.6a such that the current through M will be about 0.5 A when M conducts.

1. Adjust V_2 to 5 V
2. Keep S_C in position 'a'.
3. Trigger SCR (M) by closing S_g and adjusting R_g. After triggering SCR, open S_g.

4. Throw S_c to position 'b' and check whether M turns off or not.
5. If M is turned off, open S_c.
6. Reduce R_L to increase current.
7. Repeat steps 2 to 6 until the commutation fails. Note the maximum current that is commutated and the voltage V_3 (i.e) V_i.
8. Repeat step 2 to 7 for different initial voltage Vi.
9. Repeat steps 1 to 9 for different values of capacitor.

Calculation

The expression for capacitor current i_c is given by,

$$i_c = \sqrt{\frac{C}{L}} \sin \omega t$$

From Fig A1.6c

At
$$t = t_1 \quad i_c = I_{com}$$

$$I_{com} = V_i \sqrt{\frac{C}{L}} \sin \omega t_1$$

$$t_1 = \frac{1}{\omega} \sin^{-1} \frac{I_{com}}{V_i} \sqrt{\frac{L}{C}}$$

Where
$$\omega = \frac{1}{\sqrt{Lc}}$$

The approximate expression for the reverse bias period can be written as

$$t_{rev} = \left[\frac{\pi}{\omega} - 2t_1 \right]$$

$$= \frac{1}{\omega} \left[\pi - 2 \sin^{-1} \frac{I_{com}}{V_i} \sqrt{\frac{L}{C}} \right]$$

The reverse bias period offered by the commutating circuit t_{rev} can be computed using the above equation.

Graph

Plot the following

1. Maximum commutating current I_{com} versus the initial voltage across the capacitor for different capacitor values.
2. Computed reverse bias period t_{rev} versus the initial voltge across the capacitor for different capacitor values.

Appendix II
Short Questions and Answers

QUESTIONS AND ANSWERS BANK-I

1. Thyristor is a current controlled power semiconductor device, state true (or) false.

   Ans: True

2. How do you control the conduction of a power transistor?

   Ans: By controlling the width of base drive, we control the conduction of a power transistor.

3. What is the typical value for the on stage voltage drop of a power MOSFET?

   Ans: Less than 1 volt.

4. How a thyristor can be protected against excess di/dt?

   Ans: By using a small di/dt inductor in series with the thyristor, the device can be protected against excess di/dt.

5. Draw a half controlled thyristor rectifier.

   Ans: $V_s \rightarrow$ Supply voltage
   $V_o \rightarrow$ output dc voltage

   $V_o = \dfrac{V_m}{2\pi}(1 + \cos \alpha)$

   $V_m = \sqrt{2}\, V_s$

   α = Firing angle

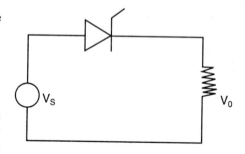

6. What is the difference between half controlled and fully controlled thyristor bridge converters?

   Ans:

Half controlled	Fully controlled
It uses one thyristor	It uses four thyristors
One quadrant converter	Two quadrant converter
At a time one thyristor is conducting.	At a time two thyristors are conducting.
Mean dc O/P voltage	
$V_o = \dfrac{Vm}{\pi}(1 + \cos \alpha)$	$V_o = \dfrac{2V_m}{\pi} \cos \alpha$

7. What is meant by continuous current operation of a thyristor converter?

   Ans: Continuous current means always the output load current is greater than zero. There is no break in the current.
   If the load inductance is too high, current becomes continuous.
   If load inductance is lesser, load current becomes discontinuous.

8. Dual converter is a four quadrant converter. State true (or) false.

Ans: True

9. What is the purpose of freewheeling diode in the case of a dc chopper?

Ans: A freewheeling diode is used to provide path for current when the thyristor switch is OFF, thereby avoiding high voltages across switch.

10. What is meant by time control ratio (duty ratio) of a dc chopper?

Ans: The ratio of on period to the total time period is known as time control ratio (or) duty ratio and it is given by

$$\delta = \frac{T_{on}}{T_{on} + T_{off}}$$

$$T_{on} + T_{off} = T$$

$$V_o = \alpha V_s$$

$V_s \rightarrow$ DC supply voltage
$V_o \rightarrow$ Variable dc supply

By varying δ (duty ratio), The output voltage can be varied.

11. Draw the circuit of Transistorised dc chopper.

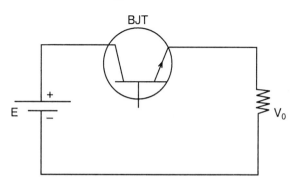

12. Give an expression for the input power factor of a single phase ac voltage controller with resistive load.

Ans: Power factor $\lambda = \dfrac{V_r I}{VI}$

$$= \frac{V_r}{V}$$

$$\lambda = \frac{1}{\sqrt{\pi}} \left(\pi - \alpha + \frac{\sin 2\alpha}{2} \right)^{1/2}$$

α = firing angle

13. What is the difference between on-off control and phase control in the case of an ac voltage controller?

Ans: on-off control is meant for any type of input signal whereas phase control is meant for sine wave alone.

In on-off control, thyristor switch connects the load to the ac source for a few cycles of input voltage and disconnects it for another few cycles.

In phase control, thyristor switches connect the load to the ac source for a portion of each cycle of input voltage.

14. Draw the circuit of a 3 ϕ half wave ac voltage controller.

Ans:

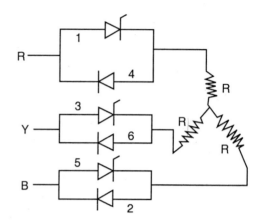

15. Why thyristors are not preferred for inverters and choppers?
    Ans: Inverters operating at high frequencies require fast acting switch. Thyristor is not preferred since it can operate at a low switching frequency of 2 kHz. Thyristors require forced commutation circuit.
16. How the output frequency is varied in the case of an inverter?
    Ans: Output frequency is varied by varying switching frequency.
17. Draw the circuit of a series inverter.
    Ans:

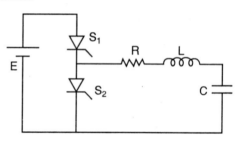

18. Give one reason for the use of microprocessors in the generation of control pulses.
    Ans: Microprocessors offer high flexibility. They can can also perform fault diagnostics
19. Give one advantage of static circuit breakers over conventional circuit breakers.
    Ans: Arc quenching mechanism (or) Arc extinction is eliminated in case of static circuit breakers. No restriking voltage occurs due to transients in static circuit breaker.
20. For what type of loads, UPS is preferred?
    Ans: For Supplying critical loads such as computers used for controlling important processes, medical equipment, air line reservation systems and hospitals, UPS is preferrred.
21. State one difference between thyristor and triac.
    Ans: Thyristor can conduct in only one direction (positive), but triac conducts the current in both directions (positive and negative). Thyristor is a three terminal device. Triac is a three terminal device and it has two SCRs connected in antiparallel inside. In order to turn on the thyristor, a positive voltage is applied between Gate and cathode. In triac, a low power trigger pulse of either polarity between the gate and main terminal M, is applied to trigger it.

22. Mention one advantage of power MOSFET over power transistor.
    Ans: Power MOSFET has very high switching speed compared to power transistor. In power MOSFET, no secondary break down occurs as in the case of BJT. MOSFET has high input impedance.
23. Mention the unit of dv/dt ratings of thyristors.
    Ans: The unit of dv/dt rating is volts/micro sec.
24. What is the main difference between half wave and full wave rectifier circuits?
    Ans: In half wave circuit, we get the output only during the half cycle. In case of full wave rectifier, the output is obtained during both the half cycles. In half wave rectifier, a pulse is obtained during the half cycle. Hence it is called as one pulse converter. In full wave circuit two pulses are obtained for full cycle. Hence it is called as two pulse converter.
25. Give the expression for average load voltage in the case of a single phase half controlled thyristor bridge converter (Semi converter).
    Ans: Average dc output voltage
    $$V_o = \frac{Vm}{\pi} (1 + \cos \alpha). \text{ (Volts)}$$
26. What is a dual converter?
    Ans: Dual converter has two fully controlled converters connected back to back. As the name implies, the dual converter has two converters. One acts as a rectifier with $\alpha < 90°$ and the other acts as inverter with firing angle $\alpha > 90°$
27. What is the effect of load inductance on the load current wave form in the case of dc choppers?
    Ans: If load inductance is high, it will reduce the ripple in the output current wave form. Load current becomes continuous.
28. Draw the typical load voltage wave form of a 1ϕ ac voltage regulator (controller) with inductive load.
    Ans:

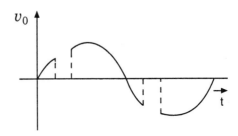

29. State the necessity of return current diodes in inverter.
    Ans: If the load connected to the inverter is inductive, then high voltages occur across the thyristor when it is switched off. This will damage the thyristor if there is no path for diverting the current. Feedback diodes (or) return current diodes are connected across each thyristor in antiparallel. The feed back diode which is connected across the thyristor feeds the reactive power to the supply.
30. State one application for series inverter.
    Ans: (i) Variable speed ac motor drives
    (ii) Induction heating
    (iii) Standby power supplies
    (iv) UPS

31. Mention one advantage of current source inverters over voltage source inverters

    Ans: (i) Commutation circuit is simple since it requires only capacitors.
    (ii) Slow response to load changes
    (iii) No feedback diodes are required.
    (iv) Utilise thyristors very effectively
    (v) Inherent regeneration feature

32. What decides the magnitude of the control pulse in the case of a power transistor.

    Ans: Load current decides the magnitude of control pulse in power transistor.

33. HVDC systems are preferred for transmitting large power over a long distance state true (or) false

    Ans: True

34. What is meant by GTO?

    Ans: GTO means gate turn off thyristor. This is a special type of thyristor. Like the thyristor, the GTO can be turned on by a short duration gate pulse, and once in the on-state, the GTO may stay on without any further gate current. However, unlike the thyristor, the GTO can be turned off by applying a negative gate cathode voltage and therefore causing a sufficiently large negative gate current to flow.

35. Triac is a dc switch. State true (or) false.

    Ans: FALSE

36. Power MOSFET is a voltage controlled device. State true (or) false.

    Ans: True

37. What is the necessity of parallel operation of thyristor?

    Ans: Whenever high current rating is needed, a single thyristor will not suffice the requirement. When two (or) more thyristors are connected in parallel, they maintain same voltage across them, but current is different through the parallel connected SCR. Therefore, we get high current rating due to the parallel connection of SCRs.

38. A fully controlled thyristor converter can function only as a rectifier. State true (or) false

    Ans: False

39. Give the expression for the input power factor of a 1ϕ fully controlled thyristor bridge converter.

    Ans: Power factor $\lambda = (2\sqrt{2}/\pi) \cos \alpha$

40. A dual converter consists of two converters connected in series. State true (or) false.

    Ans. False

41. Draw the typical output voltage wave form of a 1ϕ ac voltage controller with resistive load.

    Ans:

    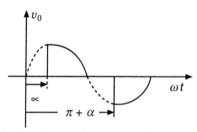

    In resistive load, voltage and current are in phase.

42. Thyristors are not preferred wherever the input is d.c. state true or false.

    Ans: True

43. How to change the output voltage of a square wave inverter?

    Ans: By using the pwm (pulse width modulation) technique, the output voltage of an inverter can be varied.

44. In a parallel inverter, capacitor is used for commutation, state true (or) false.

    Ans: true

45. What is the necessity for isolation between power circuit and control circuit?.

    Ans: It is preferred to have seperate grounds for power circuit and control circuit. If the ground of power circuit is connected to the ground of control circuit, there is no isolation. If a wire of power circuit touches control circuit, large current flows through the control circuit and it gets damaged. Hence isolation is required between power circuit and control circuit.

46. What is meant by HVDC system?

    Ans: HVDC means High voltage dc transmission system. At the sending end of power station, the generated ac power is converted into dc using controlled rectifier. Similarly at the receiving end, the dc power is converted into ac using static inverters.

47. UPS is normally used for critical loads. State true (or) false.

    Ans: True

48. State one disadvantage of static circuit breakers over conventional circuit breakers.

    Ans: Static circuit breakers cannot be used for very high currents because of the limited current carrying capacity of thyristors.

49. What is the main difference between thyristor and a GTO?

    Ans: We cannot apply negative gate cathode voltage for a thyristor to turn it off. But GTO can be turned off by applying negative gate cathode voltage and hence large negative current flows in a GTO during turnoff period.

50. Draw the snubber circuit used for dv/dt protection of thyristor.

    Ans:

    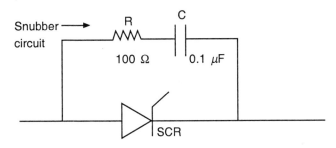

    series R–C circuit connected in parallel with SCR is called snubber.

51. Why is insulated gate bipolar transistor (IGBT) popular now-a-days?

    Ans: IGBTs have some of the advantages of the MOSFET, the BJT, and the GTO. Similar to the MOSFET, the IGBT has a high impedance gate which requires only a small amount of energy to switch the device. Like the BJT, the IGBT has a small on-state voltage. IGBTs have turn-on and off times of the order of 1 ns ie, switching speed is very high. Because of the above stated provisions, the IGBTs are populater now a days.

52. Draw the power circuitry of a 1ϕ half wave thyristor bridge converter with resistive load.

Ans:

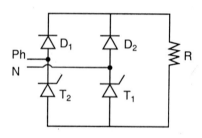

$$V_o = \frac{V_m}{2\pi}(1 + \cos \alpha)$$

53. Fully controlled thyristor converters are two quadrant converters. State true or false.

Ans: True

54. What is the input power factor of a fully controlled thyristor converter at $\alpha = 90°$ assuming ripple free load current?

Ans:

$$\text{Power factor, } \lambda = \frac{2\sqrt{2}}{\pi} \cos \alpha$$

if $\alpha = 90$

$$\lambda = \frac{2\sqrt{2}}{\pi} \cos 90°$$

$$\lambda = 0$$

55. Draw typical load voltage wave form of a voltage fed inverter.

Ans:

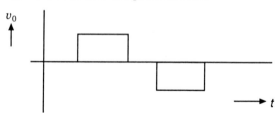

Output waveform is may be a square wave or quasi square wave.

56. What is the difference between return current diodes and freewheeling diodes?

Ans:

Free wheeling diode	Return current diode
A single diode is used.	Many diodes are connected across the thyristors.
Connected across the load.	Connected across each thyristor in antiparallel.
If provides path to ensure continuous conduction.	It feeds the reactive power to the source.

57. What is a cycloconverter?

Ans: It is a circuit which either steps up or steps down the frequency without change in voltage using one stage conversion.

58. Draw the output current wave form of a dc chopper.

Ans:

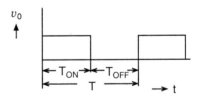

59. What is the control pulse in the case of an IGBT?

Ans: Rectangular current pulse is the control pulse in case of IGBT.

60. Draw the circuit arrangement of a 1ϕ static circuit breaker.

Ans:

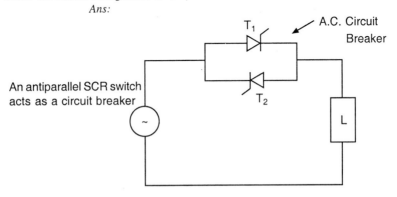

QUESTIONS AND ANSWERS BANK - II

Rectifiers

1. In case of $1-\phi$ converters, PIV rating of SCR will be highest for ...

Ans: Single phase centre taped configuration.

2. The most suitable gate triggering signal for SCR is ...

Ans: High frequency pulse train.

3. What is overlap in converter operation?

Ans: In $1-\phi$ full bridge converter, 4 SCR s conduct at a time. In centre taped type, two SCRs conduct. In six pulse converter, 3 SCRs conduct during overlap. The output voltage is zero during overlap. Overlap is due to the source reactance present in the system.

4. A $1-\phi$ bridge rectifier can operate without an isolating transformer.

Ans: Yes

5. What is the necessity of pulse transformer in the output stage of thyristor triggering circuit?

Ans: To provide isolation between power circuit and control circuit.

6. Explain pulse amplifier circuit used in the output stage of thyristor triggering?

Ans: The transistor is controlled by the pulses from the control circuit. The height of the pulse and the current supplied by the control circuit are not enough to drive the gate of SCR. The transistor amplifies the current level to the required value. When pulse is applied to the base of transistor, it conducts and acts as a closed switch. The voltage across the primary is V_{cc}. When no pulse is given, the transistor does not conduct. This circuit is called pulse amplifier circuit.

7. What is the range of α for a fully controlled converter for inverting operation? What should be the nature of load for inversion to be possible?

   Ans: (i) α between 90° and 180° (ii) The load should be a d.c. motor load to have regeneration.

8. Where FWD is needed?

   Ans: It is required for converters with R–L load.

9. What is the difference between symmetric and asymmetric semiconverter configurations?

   Ans: In symmetric semiconverter, each leg contains one SCR and one diode. In asymmetric semiconverter, one leg contains two SCRs and the other leg contains two diodes.

D.C. to D.C. Choppers

1. D.C to D.C chopper can - - - -

   Ans: Step-up and Step-down d.c voltage

2. A type-A chopper - - - -

   Ans: Always requires forced commutation.

3. Blank periods between adjacent pulses of chopper circuit are adjusted from the consideration of - - - -

   Ans: Duty ratio

4. How forced commutation methods are classified?

   Ans: Class A, Class B, Class C, Class D and Class E.

5. What is voltage commutation of thyristor chopper?

   Ans: In voltage commutation, a precharged capacitor pumps large current through the conducting SCR and turns it off.

6. What do you mean by continuous and discontinuous load current in type: A chopper?

   Ans: If the load inductance is very high and off period is lesser, the load current remains continuous. If the load inductance is lesser and off period is longer, the current reduces to zero during T_{off}. There is a discontinuity or break in the current wave form. This is called discontinuous conduction.

7. What is PWM voltage control of output of a chopper?

   Ans: In PWM, the frequency is kept constant and output voltage is varied by varying T_{on} and T_{off}.

8. To generate gate signal 555 is used in- - - -

   Ans: Monostable mode.

Inverters

1. What are the differences between VSI and CSI?

   Ans

VSI	CSI
1. Uses voltage source	1. Uses current source
2. Uses current commutation	2. Uses voltage commutation
3. 4 SCR s in each leg	3. 2 SCR s in each leg
4. Diodes are connected in parallel	4. Diodes are connected in series
5. Can be used for 120° and 180°	5. Can be used for only 120°
6. Output voltage is quasi square wave	6. Output current is quasi square wave.

2. The triggering signals for series inverter are usually derived from Q and $\overline{Q}$ of a flipflop to - - - -

   Ans. Avoid simultaneous triggering

3. What are the limitations of series inverter?
    Ans. (i) It cannot be used for low frequency.
          (ii) Output frequency has to be less than resonant frequency
          (iii) Pure sinewave cannot be obtained due to dead band.
4. Define harmonic distortion factor.
    Ans: Harmonic distortion factor is the ratio of harmonic voltage to the fundamental voltage.
5. Time margin for series inverter ensures- - - -
    Ans: Safety of devices
6. What is the difference between voltage commutation and current commutation?
    Ans

Voltage commutation	Current commutation
1. In voltage commutation, precharged capacitor pumps large current through SCR and turns it off.	1. In current commutation, L–C circuit pumps smooth current through conducting SCR and turns it off.
2. Capacitor applies reverse bias to the outgoing SCR	2. Diode applies reverse bias to the outgoing SCR.

7. What is the necessity of PWM within an inverter?
    Ans: To control voltage and harmonics PWM is used. In PWM, output voltage is controlled by varying the width of pulse. Precentage harmonics can be controlled by changing number of pulses per half cycle.

APPLICATIONS

1. Why operating frequency of SMPS is made high?
    Ans: To reduce the size of filter and transformer
2. Induction heating is done for _____
    And: Metals
3. For inducting heating, the series inverter operates at _____
    Ans: High frequency
4. HVDC transmission system with converters is economical _____
    Ans: Above 800 km
5. In 3 phase bridge converter used for HVDC application, why number of SCRs in each arm is very high?
    Ans: To meet large voltage rating, number of SCRs are connected in series
6. Dual converters when operated with circulating current have _____
    Ans: Fast response
7. What are the main components required for UPS?
    Ans: Rectifier, Inverter and Battery
8. Slip power recovery scheme refers to control of _____
    Ans: Slip ring induction motor
9. Regenerative braking of d.c motor may be achieved by _____
    Ans: Phase controlled full converter
10. Can thyristor switched capacitor provide UPF for all operating points of variable lagging pf load?
    Ans: No.

References

1. B.K. Bose, Power Electronics and A.C. Drives, Prentice-Hall, Englewood Cliffs, New Jersey, 1986.
2. P.C. Sen, Thyristorised D.C Drivers, New York, Wiley Interscience, 1981.
3. C.W. Lander, Power Electronics, McGraw-Hill Book company, London, 1993.
4. P.S. Bimbhra, Power Electronics, Khanna Publishers, Delhi, 1991.
5. S. Rama Reddy, Investigations on quasi-resonant converter fed DC drives, Ph.D thesis, Anna University, 1995.
6. V. Subrahmanyam, Thyristor control of electric drives, Tata McGraw-Hill, New Delhi, 1988.
7. G.K. Dubey, S.R. Doradla, A. Joshi and R.M.K. Sinha, Thyristorised Power Controllers, Wiley Eastern, New Delhi, 1987.
8. B. Ilango, Power Electronics, Lecture notes, Anna University, Chennai, 1985.
9. S. Rama Reddy, Applied Electronics Laboratory Manual, Madras University, Chennai, 1998.
10. S. Rama Reddy, Power Electronics Lecture notes, Madras University, Chennai, 1998.
11. G. Bhawaneswari, Power Electronic Instrumentation, Lecture notes, Anna University, Chennai, 1995.
12. S. Rama Reddy, Short questions and answers in power electronics, Madras University, 1999.

Index

A.C. Chopper using triac   110
A.C. Chopper with R load   111
A.C. Chopper with RL Load   113
Advantages of zero voltage switching   147
Analysis of IPMRI with R-load   154
Appendix-I   168
Appendix-II   178

Basic principle   67
Bipolar junction transistor   6
Bridge inverters   88
Braking of induction motor   120
Braking of dc motors   126
Base drive circuit of B.J.T.   149

Chopper   67, 109
Crow bar protection   14
Commutation   30
Classification   38
Control Strategies   68
Classification of Choppers   69
Closed loop drive   122
Closed loop controlled dc drive   126
CSI fed drive   122
Current commatution   176

Diac   17
Dynamic braking   121, 126
Disadvantages of zero voltage switching   148
Design consideration   150
Deseription of base driver board   150

Extrinsic Semiconductor   2

Fully controlled rectifier using one SCR   50
Firing circuit for 3-ϕ converter   60
Firing circuit for VCC   74
Filters   102
Four quadrant dc drive   125
Fixed capacitor banks with OCBs   134

Firing scheme for 3-phase converters proposed by Huy, Roye and Perret   163
Firing Scheme proposed by S.B. Dewan   165

Gate turn off thyristor   20
Generrex excitation system of alternators   136

Half controlled rectifier with R load   42
Half controlled rectifier with RL load   43
Harmonic control   100
Heating   12
HVDV transmission   139

Intrinsic Semiconductor   1
Insulated gate bipolar transistor   20
Jones Chopper   78

Light activated SCR   23
Load compensation   133

Methods of turn on   15
Mosfet   19
Mc-Murray inverter   90
Mc-Murray bedford inverter   92
Multiple pulse width modulation   99
Microprocessor based synchronous motor drive   143

One quadrant dc drive   123
Operating principle   148

P.N. Junction   3
PN diode   4
Protection of SCR circuits   13
Performance of rectifiers   39
Parallel inverter   87
Programmable unijunction transistor   20
PWM   102
Plugging   120, 126
Parallel redundant UPS   137

190 Index

Power electronics laboratory experiments 168

Rectifiers 38
Reverse bias condition of SCR 11
Rating of thyristor 11
Reverse conducting thyristor 22
R-Triggering 25
RC-Triggering 26
RC Triggering scheme of SCR 172
Ringing Circuit 31
Rotor resistance control 118
Regenerative braking 121
Regulated power supply 128
Resonant dc link conuerter 148
Resonant switch topologies 158

SCR characteristics 168
Series and parallel operation of SCRs 14
Silicon controlled rectifier 7
Silicon unilateral switch 22
Silicon controlled switch 21
Series LC circuit 30
Single phase rectifiers 42
Semiconverter 48
Synchronized UJT triggering circuit 58
Step up chopper 80
Series inverter 85
Single PWM 98
Single phase step down cycloconverter 104
Stepup cycloconverter 105
Speed control of induction motor 115
Slip power recovery scheme 119
Speed control of dc motors 123
Switch mode power supply 128
Static var compensators 133

Shunt reactive power compensators 134
Speed control of dc motor using microprocessor 140
Speed control of induction motor using microprocessor 142
Static power conversion 147
Single phase IPMRI 150
Single phase IPMRI with R-load 152
Single phase IPMRI with R-L load 154

Two transistor analogy of SCR 9
Triac characteristics 171
Turn off methods 32
Tap changing 100
Transformer connections 100
Two quadrant dc drive 123
Thyristor controlled static on load tap charging gear 131
Thyristor switched capacitors 134
Thyristor controlled reactors 135

UJT firing circuit 170
UJT triggering 27
Uninterruptible power supply 136

Voltage commutated chopper 72
Voltage control 98, 115
V/F control 116
VSI fed I.M. drive 121
Voltage support 133
Voltage commutation 173

Welding 129

Zero current switching 157